Corvette: American Legend
1957 – Fuel Injection/283 V-8

By Noland Adams

Published by: **Cars & Parts Magazine,** The Voice of the Collector Car Hobby Since 1957

Cars & Parts Magazine is a division of Amos Press Inc., 911 Vandemark Road, Sidney, Ohio 45365

Also publishers of:
Catalog of American Car ID Numbers 1950-59
Catalog of American Car ID Numbers 1960-69
Catalog of American Car ID Numbers 1970-79
Catalog of Camaro ID Numbers 1967-93
Catalog of Chevy Truck ID Numbers 1946-72
Catalog of Ford Truck ID Numbers 1946-72
Catalog of Chevelle, Malibu & El Camino ID Numbers 1964-87
Catalog of Pontiac GTO ID Numbers 1964-74
Catalog of Corvette ID Numbers 1953-93
Catalog of Mustang ID Numbers 1964½-93
Catalog of Thunderbird ID Numbers 1955-93
Catalog of Firebird ID Numbers 1967-93
Catalog of Oldsmobile 4-4-2, W-Machine & Hurst/Olds ID Numbers 1964-91

Catalog of Chevy Engine V-8 Casting Numbers 1955-93
Salvage Yard Treasures of America
Ultimate Collector Car Price Guide
Ultimate Muscle Car Price Guide
Automobiles of America
The Resurrection of Vicky
Peggy Sue — 1957 Chevrolet Restoration
Suzy Q.: Restoring a '63 Corvette Sting Ray
How to Build a Dune Buggy
Corvette: American Legend (The Beginning)
Corvette: American Legend (1954-55 Production)
Corvette: American Legend (1956 Racing Success)
A Pictorial History of Chevrolet 1929-1939
A Pictorial History of Chevrolet 1940-1954

Library of Congress Cataloging-In-Publication Date ISBN 1-880524-30-9
Copyright 1999 by Amos Press Inc.

Printed and bound in the United States of America

Dedicated to the memory of Kenneth Robert Wiechman

This book is dedicated to the memory of Kenneth Robert Wiechman. Ken was a friend and a 1957 Corvette enthusiast. He was driving his fuel injected 1957 Corvette when a suspected drunk driver pulled out in front of him on Friday, October 22, 1993, at the age of 48.

Ken was active in the National Corvette Restorer's Society (NCRS) and the Straight Axle Corvette Enthusiasts (SACE) Clubs. He participated in all sorts of car events, and was always ready to help. He was a friend to all, and always had a kind word for everyone. Ken was simply fun to be around. He was a sportsman, golfer, and liked other outdoor activities. At his funeral, there were relatives and many, many friends: the church was overflowing.

Noland Adams

This white, fuel injected 1957 Corvette was restored by Ken Weichman many years ago. Ken sold this car, and restored the one he was driving at the time of the accident. (CSP Calendars)

Corvette American Legend in Review

We started the history of the Corvette where it began — in the mind of Harley Earl. Earl's vision was to build an American made sports car. As head of the Styling Studios at General Motors, he was in the perfect position to get the project started.

A design was developed in complete secrecy. It was assigned the code name "Opel", a perfect cover because GM sometimes did work for their European division. The GM Opel was shown to Chevrolet management, who approved the concept, and it became the Chevrolet Opel. It was almost ready to be unveiled to the public when it was named Corvette — the Chevrolet Corvette.

The General Motors Motorama opened in mid-January of 1953 at the Waldorf Astoria Hotel in New York City. Among production cars from all the GM divisions were several "dream cars." Today we would call them prototypes or concept cars.

One of the dream cars on display was the fiberglass-bodied Corvette from Chevrolet. It rotated on a display stand while the public stared at this new design. Chevrolet was overwhelmed by the public, who wanted to buy a Corvette.

Spurred on by public reaction, the president of General Motors took advantage of the situation by announcing that the Corvette would be in production by June. This was a bold step, making a fiberglass bodied car and having it ready in five and one-half months. Fabricating fiberglass panels and assembling them into a car body on a production line had never been done before.

A building in Flint was selected to assemble 1953 Corvettes. Equipment was ordered, modified, and installed. And, like the president had said, the first Corvette was driven off the line on June 30, 1953. Only 300 '53s were built on the line in Flint.

For 1954 a new, larger facility was readied in an old building in the Chevrolet assembly complex in St. Louis, Missouri. In the rush to build a production version of the Corvette dream car, the prototype's shortcomings were transferred to the production car.

Up to this time, sports cars were all built overseas. And many such cars were luxury challenged. The Corvette followed this trend, mostly because a dream car is not intended to be driven in the rain or under other harsh conditions. The windows were chrome rimmed plastic panels, stored in the trunk and installed when weather became wet. The folding top was not developed, and it leaked air and/or rain. And, a six cylinder engine with a two-speed Powerglide automatic transmission was the only combination available. These items, along with the fiberglass body and the limited color choices were meeting resistance in the sales department.

On the other front, Ford countered the Corvette with its Thunderbird. A miniature version of the 1955 Ford convertible. The Thunderbird was an instant hit with its steel body and a wide choice of colors. The Thunderbird had a V-8 engine from the beginning, a choice of transmissions and had roll-up windows in the doors.

Corvette sales in 1954 were in a slump. In spite of the addition of a V-8 in the 1955 Corvette, production was almost canceled. Things had to be done to save the Corvette, or would GM let it die.

At this point, die-hard Corvette enthusiasts were few and far between. There was no organized group looking for details on the new 1956 Corvette. Chevrolet's engineer Zora Arkus-Duntov knew the Corvette needed publicity. He prepared a Corvette mule (test car) to set a speed record. In early January of 1956 Duntov drove the mule Corvette to a two way average flying mile of 150.583 mph on the sands of Daytona Beach.

The GM Motorama opened a few days later at the Waldorf-Astoria Hotel in New York City. The record of 150 plus mph was highly publicized, and the 1956 Corvette gained the attention it needed so badly. This was the turning point in Corvette history. Chevrolet offered the Corvette with color choices, transmission selections and roll-up windows. Sales picked up, and enthusiasts began following the Corvettes as they raced.

After the January run on the beach, the Corvette mule was joined by two stock 1956 Corvettes. All three ran at the speed trials during Daytona Speedweek in February 1956.

Chevrolet now had the cars race-prepared for the sports car track at Sebring. Even though the cars had a number of problems, they did well enough to bring more attention to Corvette.

Through 1956, Corvette continued its racing schedule, and won many races. In support of its racing efforts Chevrolet engineering's group was debugging a fuel injection system designed by John Dolza. The system would eventually be manufactured by GM's Rochester division, and installed on 1957 Chevrolets (including Corvettes) and Pontiacs. Plus a larger V-8, up from 265 cid to 283 cid, was to be installed in 1957 Chevrolets.

So 1957 opened on a bright note.

Author's Introduction

1952: a Corvette prototype is built. 1953: the prototype is shown to the public; the Corvette is rushed into production. 1954: production moves to St. Louis, but sales lag. 1955: in spite of a new V-8 engine, sales are low. 1956: a body facelift, new options, and racing successes get the public's attention; sales improve.

That brings us to 1957, only the fifth year of Corvette production. The 1957 body was identical to the 1956 body. Likewise, the suspension was unchanged for 1957. Thus, the engine and transmission were the mechanical improvements for 1957.

In only its third year of production, the Chevrolet V-8 for 1957 was increased from 265 cubic inches of displacement (cid) to 283 cid. This, along with several special options and mechanical improvements, would produce a 1957 Corvette that many revere as a highlight year.

Another optional engine was equipped with the Duntov cam and heads with larger valves. But it had a significantly different induction system: fuel injection built by Rochester Products, a division of GM. The Rochester Fuel Injection system distributed fuel directly into the intake manifold runner through eight nozzles. As a result, the fuel injected Chevrolet engine was smoother and more responsive than carbureted engines in 1957.

The fuel injection system was developed by John Dolza, and refined for production cars by Zora Arkus-Duntov. The full story of its development is elsewhere in this book. The fuel injected engine was also available in full sized Chevrolet passenger cars in 1957. In dynamometer tests, the new fuel injected engine produced 290 horsepower (hp). The test was run again — and again, and on several engines. It produced a solid 290 hp. But that wouldn't get anyone's attention, so the fuel injected engine with the Duntov cam was officially rated at 283 hp. And advertisements made a big noise about one horsepower per cubic inch: 283 hp per 283 cid! Yet it wasn't enough that the 1957 Corvette was improved and ready for the general public to purchase. It needed to be brought to the attention of potential new Corvette buyers. Chevrolet had gained much needed publicity for the Corvette through racing in 1956. And 1957 promised even more racing successes.

The head of the GM Styling Studios, Harley Earl, had a racing version of the Corvette, called the SR-2, built for his son. Bill Mitchell, Earl's assistant at Styling, had a second SR-2 built for himself. A third SR-2, based on a stock chassis, was built for the President of General Motors, Harlow (Red) Curtice.

Based on his experiences in Europe, Chevrolet's engineer Zora Arkus-Duntov pushed for a special Corvette. This was to be a super special racing version, not available to the general public. Finally, in late 1956, Duntov got approval and began building the XP-64, which became the Corvette Super Sport (SS).

The SS project began with the construction of a test mule. Mules are crude vehicles used for testing systems like suspensions or transmissions. A special lightweight tube frame was built for the SS mule. A stock base Corvette engine, rated at 230 hp, was installed in the mule. The internal body supports were made of plywood. A fiberglass body was fabricated and installed on the mule.

The results of testing the mule influenced the final design of the Corvette SS. The tube frame and suspension was special, just like on the mule. But, the body was a masterpiece, hand crafted from magnesium.

Chevrolet had plans to run the Corvette SS and Mitchell's SR-2, along with two stock 1957 fuel injected Corvettes at the 1957 Sebring races. This would be followed by entering the same cars in the 24 hour race in Le Mans, France, later in the year.

The drivers test drove the white Corvette SS mule many times around the Sebring course. The SS itself was being finished by the mechanical and styling studio engineers and technicians. Since the body was built and completed by the styling staff under Harley Earl, it had to be flawless. A lot of time was spent making sure the SS's appearance was as perfect as a show car.

By the time the race version of this Corvette SS made an appearance at Sebring, there was no time left to test the car. The hours spent test driving the mule would have to do. Since they were essentially the same car, no one expected a problem. But the SS had several problems, finally withdrawing from the race. But the SR-2 finished the race!

Meanwhile, Corvette's production engineers were improving the 1957 production Corvette. After fuel injection came a four speed manual transmission. Then heavy duty suspension improvements. Then a special cold air inlet box routed cold air into the fuel injection's intake system. The Corvette was becoming a serious race car.

Then the axe fell. On June 6, 1957, the Automobile Manufacturer's Association issued a ban on racing. It meant the official end to Chevrolet's support of racing.

These are some of the highlights of the 1957 Corvette production year. We shall examine each of these subjects — and more — in detail in this book.

As always, your comments are welcome. Please contact me at: PO Box 1134, El Dorado, CA 95623.

Noland

Contents

Chapter 1: Design Proposals 3

Chapter 2: Production Begins 7

Chapter 3: AIM: Assembly Instruction Manual . 36

Chapter 4: Harley Earl Builds the SR-2 39

Chapter 5: Corvette Super Sport 74
        SS Construction 100
        SS Track Performance 146

Chapter 6: Fuel Injection 175

Chapter 7: Auto Shows 194

Chapter 8: Daytona 200

Chapter 9: Sebring 212

Chapter 10: 4-Speed Transmission 247

Chapter 11: Body Modifications 251

Chapter 12: Mid-Year Performance Changes . 254

Chapter 13: Corvette Media Origins 262

Many of the photographs in this book originated in the GM Media Archives in Detroit. As per agreement, use of those photos require this statement:
"Copyright 1978 GM Corporation, used with permission from the GM Media Archives."
Thanks must also go to the Chevrolet Communication folks, part of the Chevrolet Motor Division, who provided the photos which originated in GM Photographic or the GM Media Archives.
The author also wishes to thank several other sources of photos. The credit for those photos appear with each image.

Design Proposals

Chapter 1

The proposed 1957 Corvette shown in a full sized rendering, shaded to show detail, photographed on February 1, 1955.

The GM Styling Studios, under the direction of Harley Earl, produced new sketches for every Corvette year. There had been a major body facelift for 1956 production. Internally, the '56 Corvette was very similar to the 1955, but the body looked quite different.

In planning for 1957, Mr. Earl's studios produced several full sized drawings in 1955. There were several side trim variations. The front grille area, which had 13 vertical teeth, originated with the 1953 Motorama show Corvette. Another version had a front end based on a combination of Earl's 1951 Le Sabre show car and the 1955 Biscayne Motorama car. Both front end designs — grille teeth, Biscayne, or Le Sabre center bullet — were Mr. Earl's choices. The toothy grille design from previous years was retained, and the rest of the '57 proposed design was never developed further.

The 1957 Corvette full sized rendering with the top up. The side cove and the rear exhaust treatments are similar to 1957 production items, but not much else.

Front view of the proposed 1957 Corvette, resembling a combination of Harley Earl's 1952 Le Sabre and the 1955 Biscayne Motorama show car. Since Harley Earl liked the Carl Renner designed grille teeth, they won out over this design, fortunately.

Another Corvette proposal was this version, photographed on April 26, 1955, with finned rear fenders and vent side windows. From what can be seen, nothing was used on the 1957 production Corvette.

Production Begins

Chapter 2

These are the major body panels used to fabricate the 1957 Corvette's body.

Production of 1957 Corvettes began on Friday, Sept. 21, 1956. The last 1957 Corvette was built on Friday, Sept. 6, 1957. That's eleven and a half months to build and sell 6,339 Corvettes. This is a significant figure, we'll tell you why.

From Chevrolet's records, Corvette Production:

1956, 3,467 units

1957, 6,339 units

Now, that's quite an increase from 1956 to 1957. Or is it? Let's see, 1956 production began on Monday January 9, 1956. The last 1956 Corvette was assembled on Thursday Sept. 20, 1956. Only four 1956 Corvettes were assembled the last day. The next day, Sept. 21, 1956, the 1957 model production began and four 1957 Corvettes were assembled.

My point here is that this was a major manufacturing schedule change. Model year production had always begun at the start of the calendar year. That is, the assembly lines for 1956 Chevrolet cars and trucks started the first week in January 1956. But for 1957, the assembly plants would begin building Chevrolets four months earlier in September of 1956.

Ref: The Corvette Birthday Book

This was an industry-wide change designed to have "next year's" cars in the dealer's showrooms "this year." Production records from Chevrolet show that all Chevrolet truck and passenger car plants followed the same pattern. They all had a short production year for 1956. Under normal circumstances, there was a 2 to 4 week shutdown at each assembly plant for a model changeover period. There were new parts to arrive, new equipment, jigs, and fixtures to be installed, all necessary to assemble the new models.

This early production start was supposed to give Chevrolet an edge over their competition. Since we don't have access to records from other automobile manufacturers, we can guess that Chevrolet was: 1) Ahead; 2) Tied with; or 3) Catching up with the competition.

Just think, you could have a new next year's Chevrolet model delivered by Christmas. Let's see the Jones' keep up with that!

In the case of the Corvette, the 1956 to 1957 assembly line changeover was minimal. Probably just enough time to sweep the floors, check the alignment and condition of the various equipment and fixtures.

Here is a closer look at the schedule. Thursday, Sept. 20, 1956 was the day the last four 1956 Corvettes were assembled. The next day, Friday, Sept. 21, the first four 1957 Corvettes were assembled. Figuring the time required to assemble the bodies, prepare them for painting, prime them, send them through the paint drying ovens, sand the bodies and correct any imperfections, apply the first color coat, go through the ovens again, sand the body lightly, apply the final color coat and a last trip through the ovens! That's a lot to do, but the assembly line can do it in a minimum of time. It takes about a day and a half to build a 1956-57 Corvette body. Assembling the chassis takes much less time. Now we can visualize what happened on Wednesday, Sept. 19, 1956. As the last of the '56s started through the St. Louis line that morning, the assembly line immediately began building 1957 Corvettes. Remember, those last '56s were identical to the first '57s, so there was no problem at all.

Body panels fabricated by an outside supplier are waiting in an outside storage area until they are needed.

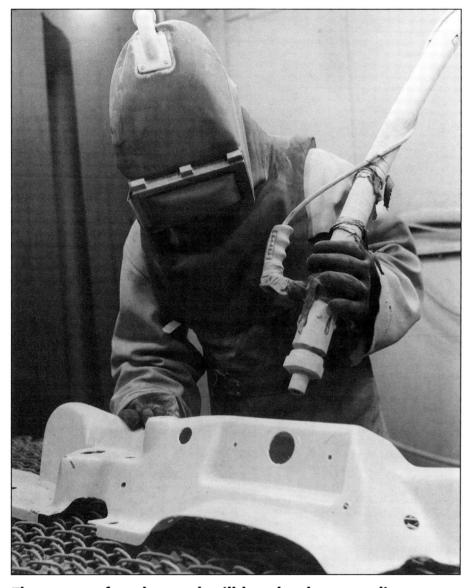

The areas of each panel will be glued to an adjacent panel, and are blasted with shot to rough up the surface so the adhesive will produce a strong joint.

Now let's make a new comparison chart.

Corvette Production:

1956, 8-1/2 months, 3,467 units, 408 per month.

1957, 11-1/2 months, 6,339 units, 551 per month.

So 1957 production quantity improved slightly over 1956 in the number of units built per month: from 408 to 551 units. On the surface, the 1956 figure of 3,467, which increased to 6,339 in 1957 looks like a 45 percent increase. But when we factor in the total production time for each year, we see an increase of only about 25 percent.

THE "NEW" 1957 CORVETTE

The 1957 Corvette was based on the 1953 Corvette in many ways. The suspension was almost identical. Chevrolet engineer, Zora Arkus-Duntov, improved handling by installing an aluminum wedge on each side of the front suspension. In the rear, Duntov lowered the rear springs front mounting position. Except for these two minor changes in 1956, the entire suspension was unchanged from the '53 model.

The body looked quite different from the 1953 to 1955 body style. There had been several changes for 1956, all of which carried over for 1957. The 1956 and early 1957 bodies were identical, so we will examine the 1956, early 1957 body.

Compared to the 1953-55 body style, the 1956 body style appeared to be entirely new. In reality, body changes were limited to the front area, sides, rear, hood,

The large floor pan is the first part to be fastened to the body dolly. Note the job number 162 on the firewall in front of the driver.

trunk, and windshield. The exterior changes for 1956 were carried over into 1957 unchanged.

Mechanically, 1957 Corvettes are nearly identical to 1956. The base transmission is the same, a three-speed manual transmission, with an optional two-speed Powerglide automatic transmission. An optional four-speed manual transmission will be added later. The rear differential is unchanged from 1956. A limited slip rear differential called Positraction will be added later.

The major change for 1957 was an increase in engine size from 265 cubic inch displacement (cid) to 283 cid. The 265 and 283 blocks were identical in appearance, but the larger 283 engine was improved in many ways. There were improvements in heads, camshafts, and intake manifolds. In addition, the maximum rated horsepower for the 1956 265 cid engine was rated at 240 horsepower. Combined with the new add-ons, the 1957 283 cid engine produced a maximum of 270 hp. Later in the production year, the fuel injected high horsepower engine would produce 283 hp. With a favorable power to weight ratio, the 1957 Corvette gained a large following of dedicated enthusiasts for the first time.

The one piece rear section is installed. Note the trunk hinges are already installed. Written in the trunk is job number 59, and PT. The two large holes in the front panel are for the power top wiring and hydraulic hoses.

CORVETTE: AMERICAN LEGEND

The rear upper body has been installed, while the front upper body waits. Note the installation of the wiper transmissions, radiator support, small clips, and hood hinges.

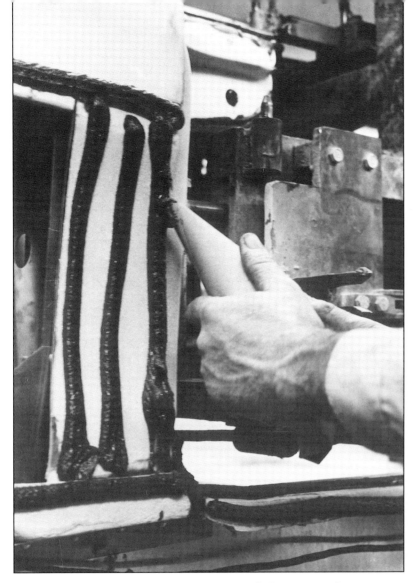

"Bond", the adhesive used to fabricate the bodies, is applied to the side cove area.

The roughest areas are smoothed with grinders.

CORVETTE: AMERICAN LEGEND   15

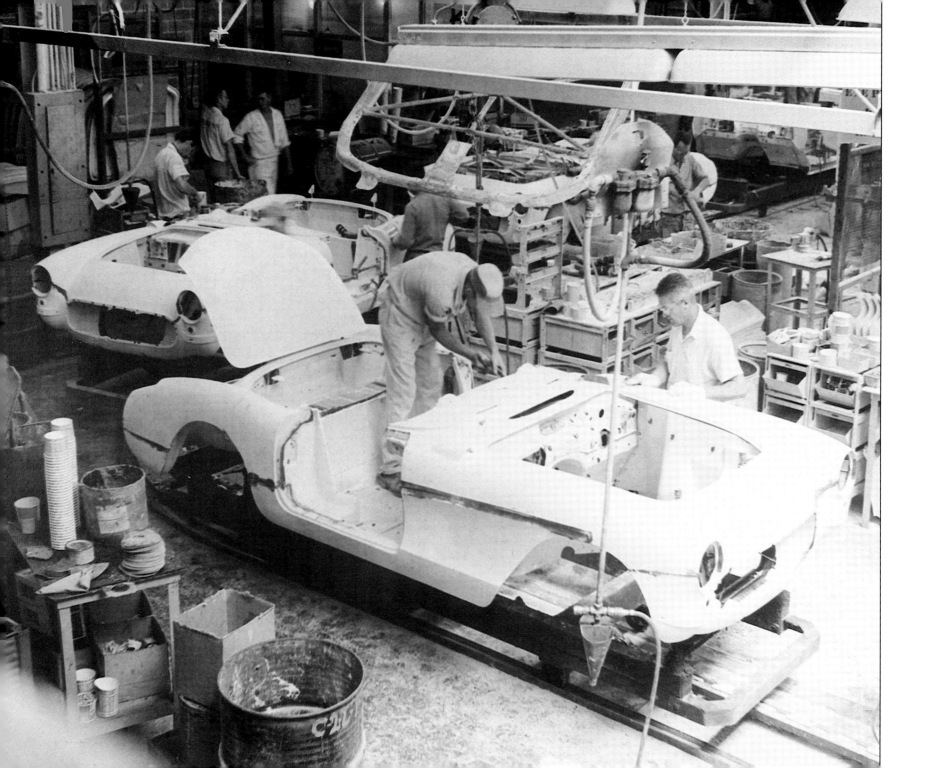

(Left) Installation of the front body section. The unpainted bodies are called "bodies in white" at this point.

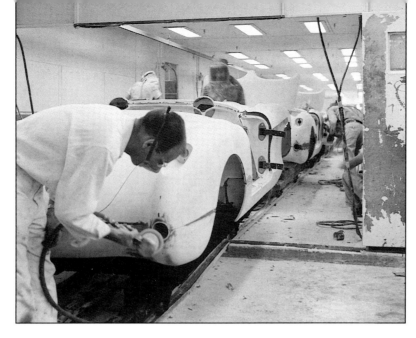

Preparation for painting begins by sanding all surfaces prior to spraying the primer coat.

The bond joints are smoothed with a grinder.

The surfaces to be painted are finished by hand. Note the rough area beside the taillight housing which must be repaired.

CORVETTE: AMERICAN LEGEND

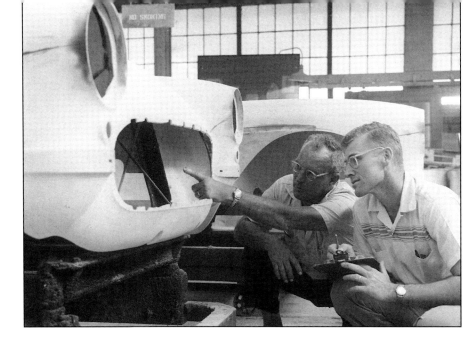

Inspection of the bodies just before entering the prime coat paint booth.

Spraying on the prime coat. This is a recirculating air booth, so masks are not necessary.

Wet sanding the prime coat with a pneumatic sander, which is also spraying water on the surface.

The prime coat sanding line.

After applying the first color paint coat each body is carefully sanded and then buffed with rubbing compound.

Primed and painted bodies coming out of the drying ovens.

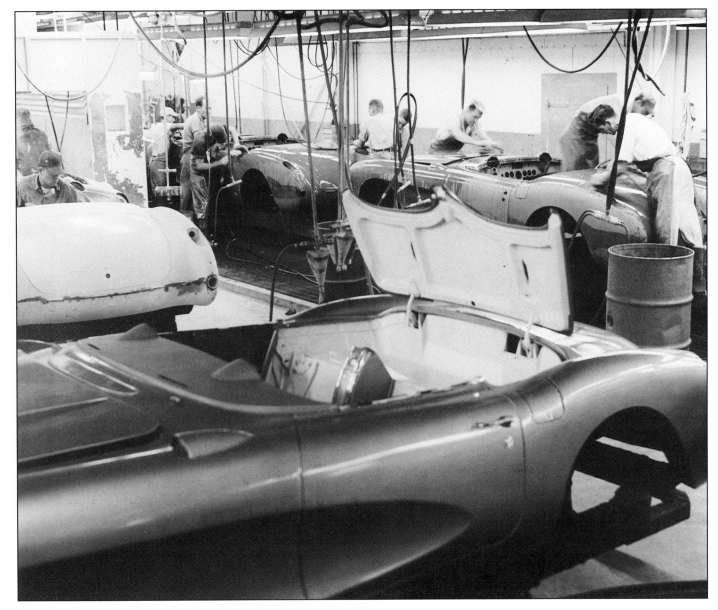

Bodies in various stages. Left, an unpainted body in white; background, the prime coat sanding line, without door handles; foreground, a painted body with a lighter interior, with door handles.

CORVETTE: AMERICAN LEGEND

Spraying on the final color coat.

Buffing out the final color coat.

Finishing touches, installing the windshield, as the body moves down the assembly line.

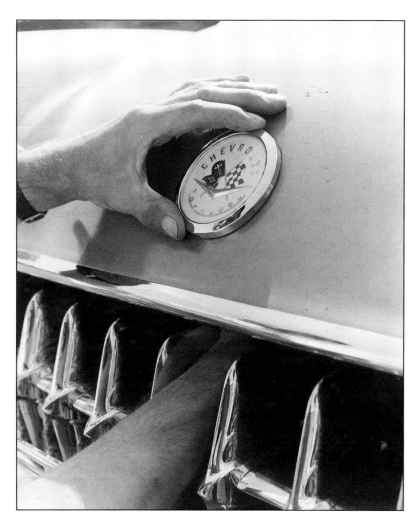

Completing the body assembly by installing the exterior trim.

Rear brake light and trim installation.

Next step is finishing the interior, here they install a door panel.

Last step in the interior is installing the instruments.

End of the line, the body is done.

CORVETTE: AMERICAN LEGEND

The final body inspection. Everything is checked and inspected before it leaves for the chassis and body assembly line.

While bodies are being built, painting, and trimmed, the much shorter chassis line has been installing mechanical components. Frames are brought inside from their outside storage area, where they are placed upside down on the chassis line. The front and rear suspension assemblies are installed, the chassis is turned over and the engine and transmission are installed as shown here.

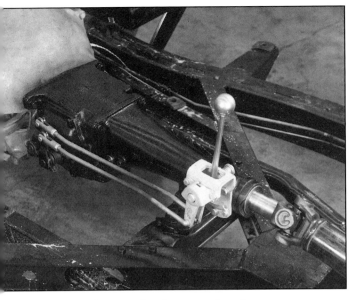

Pictured above, the base transmission in 1957, the three-speed manual.

After the engine and transmission and a few more parts are installed, the body is lowered onto the chassis. The chassis in the foreground has an engine with two four-barrel carburetors.

The base engine, rated at 220 hp, with a Powerglide transmission.

Mechanical connections are done from below. Note the body color "red" in chalk on the front crossover.

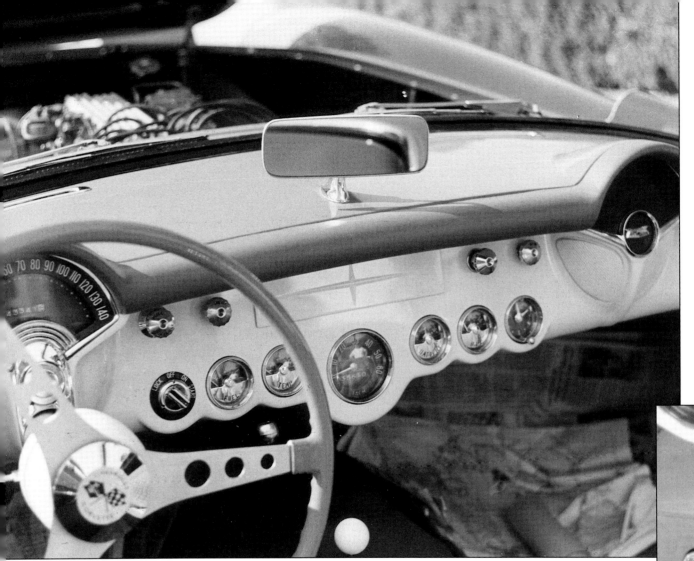

The dash instruments in a '57 without the optional radio.

The shifter in a three-speed 1957 Corvette.

32 CORVETTE: AMERICAN LEGEND

The interior in a new completed '57.

New '57 Corvettes in the shipping lot. The third car back has a protective covering over the folding top — that's not a wrinkled top.

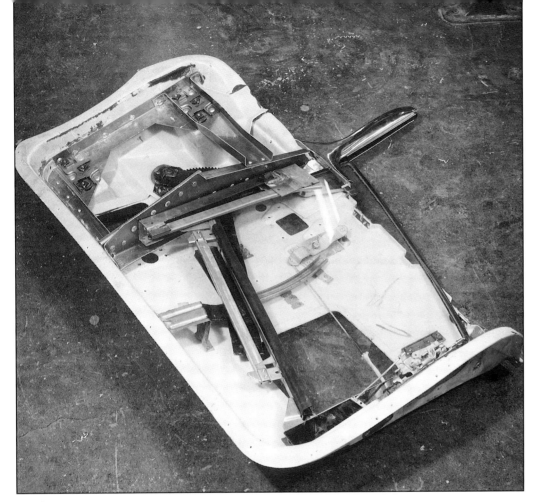

In mid-1957 production bracing was added to the doors (shown here) and under the dash.

All tagged and legal.

AIM: Assembly Instruction Manual

Chapter 3

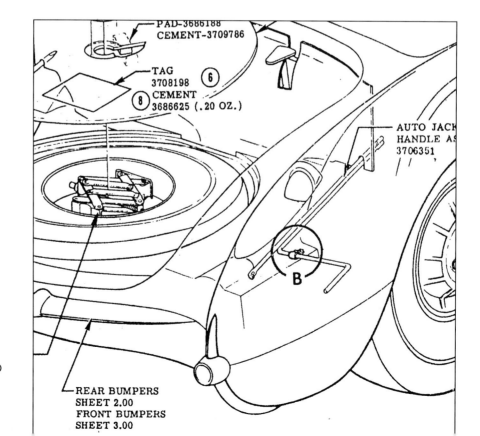

The first Assembly Instruction Manual (AIM) for the Chevrolet Motor Division was produced for the 1956 model year. Included were sedans, pickups, and Corvettes. Each model had its own unique AIM. This made production and assembly easier for the production workers.

Before the AIM, production workers used the line drawings for assembly instructions. These were large sheets that illustrated the entire layout of a section. They were difficult to handle, and the drawings were not clear. The AIM was designed to show the assembly line worker the exact details of how to assemble each part so that all the cars were properly assembled. Complex sections were shown in exploded views, with parts identified by individual parts numbers.

Another reason for the individual pages in the AIM was that it was easy to upgrade when a procedure or part was changed. As a part was improved or replaced, a new AIM page was drawn. The old page was discarded and replaced. Thus, the AIM changed along with the car to represent the current assembly line details.

The 1956 and 1957 Corvettes were so similar that a new AIM was not drawn up for 1957. The engineers just kept changing and updating the AIM pages. By the end of 1957 production, the 1956 AIM was almost entirely replaced by the revised 1957 pages. Therefore, the 1956-1957 AIM currently available is an accurate representation of only the last few weeks of 1957 production.

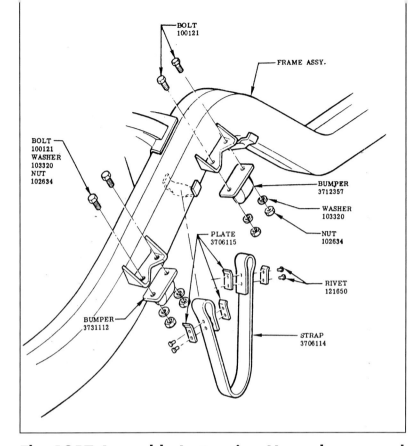

The 1957 Assembly Instruction Manual was used to illustrate the installation of important assemblies; this is the rear axle rebound strap (driver side shown) and the two rubber bumpers that limit rear axle movement.

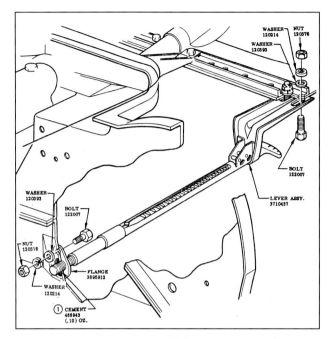

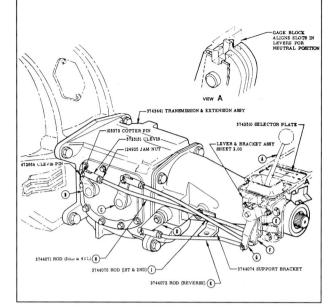

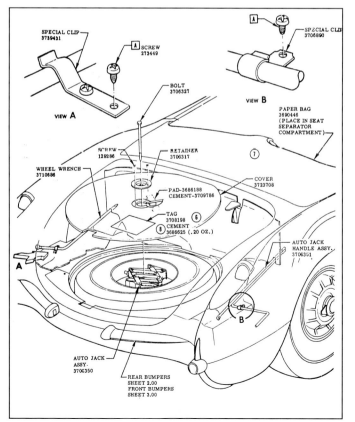

The AIM is very useful in re-installing assemblies and parts after repair or restoration; the parking brake lever is shown here.

As new options were added to the 1957 Corvette, the installation was shown in the AIM. The four speed manual transmission, Regular Production Option (RPO) 685, was added on April 9, 1957.

The installation of tools in the trunk is detailed here.

Harley Earl builds the SR-2

Chapter 4

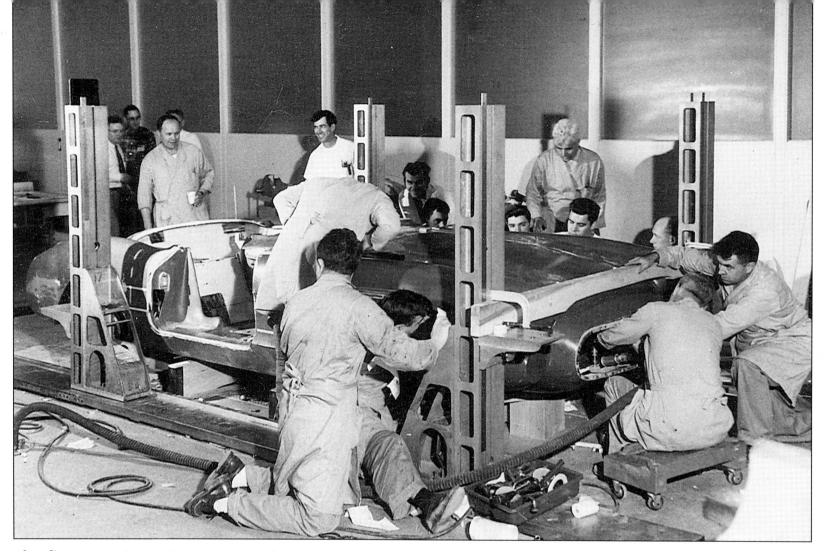

The first SR-2 is getting a nose job under the direction of Harley Earl. There are 16 or 17 engineers and technicians in the photo.

The history of the SR-2 Corvettes began in 1956 when Jerry Earl, a resident of Michigan, decided to race sports cars. He liked the Ferrari, so he bought one. This would not have been a problem, except Jerry's father was the one and only Harley Earl. Mr. Earl was the head of the General Motors Styling Studios, and a Vice President of GM. Mr. Earl's studios produced the Corvette, and many other Motorama show cars. It must have irritated Mr. Earl that his son, Jerry, wanted to race a non-GM car.

Of course Jerry was right, the Corvette, made by GM's Chevrolet Division, did not have a race-proven sports car like the Ferrari. So Mr. Earl made a deal with Jerry. I'll have the studios build you a racing Corvette, using the heavy duty components first used at the Sebring track in Florida. We'll give it distinctive styling, and you'll have a GM sports car. Then you can sell the Ferrari.

Okay, so Jerry sold the Ferrari. The styling studios brought in a stock 1956 Corvette, and modified it as directed by Harley Earl. Based on the Sebring Racers, it became the SR-2. The '56 Corvette chosen was E56S002522. Because the sequential numbering system began at 001001, this was the 1,522nd 1956 built. Its final assembly date at the St. Louis Assembly Plant was Monday, May 21, 1956.

Harley Earl's number one assistant on the GM Styling Staff was William L. Mitchell. Bill Mitchell was an experienced stylist and a trusted employee. As the first SR-2 neared completion, Mitchell decided that he wanted an SR-2 as well, and he got it. Mitchell's 1956 corvette was serial number E56S002532, only 10 serial numbers after the first SR-2 that was to be built for Jerry Earl. Mitchell's number 1,532 was built in St. Louis on Tuesday, May 22, 1956.

But Mitchell's SR-2 was different. There were crossed flags in the side coves, and a real nose emblem. The first SR-2 had a decal instead of a nose emblem. But the really big difference was the rear fin. Instead of being rather low and placed in the center of the trunk area, it became a streamlined headrest, covering the roll bar and the fuel filler neck and cap.

A third SR-2 was to be built for Harlow "Red" Curtice, the President of General Motors. Unlike the other two SR-2s, Curtice's car was built for show. It had no chassis or brake modifications, just special appearance items. Curtice's Corvette was E56S002636, the 1,636th 1956 Corvette. It was assembled in St. Louis on Friday, May 25, 1956.

There are only 114 serial numbers between the first SR-2 (002522) and the third SR-2 (002636). All were brought to the mechanical shop of the styling studios. There they were assigned shop order numbers, a way of tracking labor and materials costs. The shop orders assigned were:

E56S002522   S.O. 90090

E56S002532   S.O. 90100

E56S002636   S.O. 90101

The Jerry Earl SR-2 had the highest priority, no cost or manpower was spared. First, drawings were produced by Harley Earl's styling studios showing the modifications to be made to the body. This included designing, fabricating, and installing a new nose, modifying the folding top lid cover and trunk lid, preparation and painting, then installation on the modified chassis. The force behind this project was Harley Earl himself: perfection would be a minimum requirement. To achieve this goal in such a short time, as many as 17 engineers and technicians at a time were seen working on the Earl SR-2 project.

The front end design of the Corvette Impala Motorama show car as modified.

Earl's 1956 Corvette was assembled by the St. Louis Corvette Plant on May 21st. By May 31, 1956, the body showed considerable changes: the trim and most of the paint had been removed. The body was ready to be removed from the chassis, and both the body and chassis were prepared for modifications.

Styling the front of the SR-2 was easy. Harley Earl wanted to retain the "trademark" Corvette grille teeth and outer grille shell. The styling staff used the design from the 1956 Corvette Impala dream car which was built for the 1956 GM Motorama shows. A new lightweight extended nose section was fabricated from fiberglass. The stock front end was removed and the new nose was spliced in place.

Meanwhile, all the heavy duty optional parts that had been designed and fabricated for the Sebring cars were being installed on the chassis of E56S002522. These included stiffer shocks, a larger front stabilizer bar, a quick-steering adapter, heavier springs front and rear, cermatallix brake pads, and Halibrand wheels.

By June 13, the painted body was having its trim installed. On June 15 the car was ready, complete with the SCCA assigned number 144. In three weeks a stock 1956 Corvette, serial number E56S002522, was turned into the impressive SR-2.

That brings us to a mystery that has not yet been solved. The first SR-2 Corvette, as we said, was E56S002522. This number was written on the rear of the frame in crayon, as seen in photos of the completed chassis. Yet the serial number plate and the registration papers have the S replaced by F: E56F002522, indicating a 1956 Corvette built in Flint, Michigan. The stock 1956 Corvette, indicated by "S" in the serial number, was built in St. Louis, Missouri. It was modified

The front emblem on the 1956 Corvette Impala.

into an SR-2 in the mechanical shops of the Styling Studios at the GM Tech Center in Warren, Michigan. The presence of the "F" in the first SR-2's serial number plate remains a mystery.

Earl's SR-2, painted a light metallic blue, was entered in the SCCA races at Road America in Elkhart Lake, Wisconsin, June 23-24th. This was barely a month after it was built in St. Louis as a stock 1956 Corvette. Wearing its low fin and number 144, the car was driven at Elkhart Lake by Jerry Earl. The car was heavy and hard to handle, and Jerry spun out at least once. Dr. Dick Thompson, a well-known Corvette driver was there, and he took over for the balance of the six-hour race, finishing in a respectable position. He found the car to be

The 1956 Corvette with the rear fin in place, showing the molds for the two small windscreens.

overweight. Weight reduction with increased horsepower was needed.

One report on the SR-2 running at Road America: "Modified Corvette number 144 got off to a very bad start and was in the pits for over six minutes. The SR-2 finished 16th overall but our watches recorded lap times of 3:05 throughout most of the race, as compared to 3:08 to 3:10 lap times for the best placed Corvette which finished 10th overall." — *Road & Track*, September 1956

Back in the Styling Studios, the stock deluxe interior, including door panels, blue leather seats, all the glass and internal door parts — except the door opening hardware — was removed. The blue leather transmission tunnel cover and the tunnel mounted fire extinguisher remained unchanged. The replacement door panels were thin concave, lightweight fiberglass panels. Lightweight seats from a Porsche replaced the original Corvette seats. The weight reduction program brought the overall weight down by 300 lbs.

With Earl's SR-2 out of the mechanical shop, work began on Mitchell's SR-2. It was similar in appearance, but it had several variations. In the front, a large emblem was mounted above the center of the grille. This emblem became the prototype for the 1958 Corvette. Earl's SR-2 had an emblem-like decal on the nose, not a true emblem.

A crossed flag emblem, much larger than emblems on stock 1957 and later Corvettes, was installed in the side cove. Earl's SR-2 had a crest-like emblem with wings in the side cove.

Mitchell's SR-2 was fitted with a huge headrest/fin. From a side view it looked similar to the headrest/fin that had been installed on Duntov's test mule in December of 1955. Unlike the 1955 add-on fin, the SR-2 fin was hollow and built into the trunk lid. Also, the fin concealed the 36-gallon fuel tank filler neck and cap.

Mitchell's number two SR-2 was completed in early September. Both cowl scoops had their internal parts removed, making them functional air scoops. It had small twin windscreens instead of a single windshield. The tires were special, with extremely narrow white walls. In the front the parking lights are in place, replaced by screens later.

Mitchell was a flamboyant person. After test runs at Road America and Marlboro, Maryland, he entered the SR-2 in the Daytona Speedweeks Performance Trials. The car got a special red and white-trimmed paint job, with the fin painted in red and white stripes like an old fighter plane. It had a hinged canopy, again like an old fighter plane. The headlights had streamlined fairings, with flat "Moon" discs on the front wheels, and skirts on the rear. Having previously been equipped with two four-barrel carburetors, the 336-cid engine was fitted with fuel injection. Fuel injection emblems were added to each side cove under the large crossed flags.

The Daytona Beach race began on February 7, 1957. In those days the schedule was dictated by the tides. The driving chores were shared by Betty Skelton and Buck Baker. Buck Baker won the modified class with a standing mile average of 93.047 mph. In the flying mile, its two way average was 152.866; only a D-type Jaguar was faster.

Next was Sebring, where the car was repainted red with white trim, but without fin stripes. Despite various mechanical problems, it finished the race in 16th place. More details of this SR-2's racing history are in the specific sections on Daytona and Sebring in this book.

Meanwhile the chassis has had all the heavy duty racing parts installed.

Heavy duty parts: the front brakes and the stabilizer bar.

After the AMA ban, the Mitchell SR-2 was retired. It sat in a basement on General Motors property until it was sold. It is currently owned by Bill Tower of Florida, who has restored it to excellent condition. It can be seen at auto shows throughout the U.S.

Harley and Jerry Earl liked the large tail fin on Mitchell's SR-2. They ordered a large headrest/tailfin to be fabricated and installed on the original SR-2. Work had been started on the third SR-2, and it had an instant low fin!

Jerry Earl raced the SR-2 number one, now with a large fin, at several other races. It was seen at Road America in June 1957 wearing number 100. In the fall of 1957, Dick Thompson drove it to victory at Marlboro, Maryland. In November 1957 it was entered in the Nassau race with Curtis Turner driving. The car ran well, but did not finish at Road America or Nassau.

After Nassau, Earl's SR-2 was sold to Nickey Chevrolet in Chicago. It was painted purple, and became known as the Purple People Eater! Nickey handled the mechanical work and Jeffords did the driving. The result was a national championship in SCCA in B production in 1958.

After passing through several owners, it is now owned by Rich and Char Mason. The car is now restored to good-as-new condition. And it is not just a show car, Rich drives it in vintage sports car races. This author had the honor of driving SR-2 number one for several slow blocks a couple of years ago.

Now back to SR-2 number three. The low fin, side cove scoops, extended nose, and special lights were also incorporated into this otherwise stock Corvette, which became SR-2 number three. It was fitted with handsome chromed Dayton wire wheels. This was for the use of the GM President Harlow "Red" Curtice, and was finished in the special metallic blue used for all his family cars.

As was the case with the Earl SR-2, the interior was blue leather. The exterior of the hardtop was covered by a sheet of stainless steel. There were other details added that were fitting for the President of General Motors.

After Mr. Curtice drove his SR-2 for a while, he sold it to a friend. It had several more owners until it ended up on a used car lot in rural Michigan. Richard and Carylin Fortier, owners of Paragon Reproductions in Swartz Creek, Michigan bought the car. It is in excellent condition, and this author had the privilege of riding in this SR-2 several years ago.

The painted body has been re-installed on the chassis. Interior and exterior parts are being installed on June 13, 1956.

The finished SR-2 with Jerry Earl in the driver's seat with the Chevrolet Studio Head Stylist Clare "Mac" MacKichan as passenger.

The SR-2 posed in the Styling viewing yard at the Tech Center.

Chevrolet General Manager Ed Cole in the number one SR-2.

Dr. Dick Thompson driving Earl's SR-2 at Elkhart Lake.

Mitchell's number two SR-2 with cowl scoops and parking lights, just completed on September 4, 1956.

The Mitchell SR-2 was raced several times before Daytona. Note that the cowl scoops are in place, and so are the parking lights. Also, the car is carbureted at this point, and there is no side cove fuel injection emblem.

Bill Mitchell with his SR-2 on February 4, 1957. The cowl scoops have been removed, a hinged bubble canopy and a cover for the passenger's compartment have been added. The parking lights have been removed and covered with streamlined fairings to match the headlight treatment. Fuel injection and fuel injection emblems in the side cove have been added, and the car has an eye-catching red and white paint job for the races at Daytona.

Mitchell's mighty SR-2 is unloaded on the beach at Daytona.

Betty Skelton perches on the SR-2 for a publicity shot. (Daytona Speedway Archives Photo)

Betty Skelton makes a low speed publicity run for the cameras at Daytona Beach.

The Mitchell SR-2 poses in the Sebring pits with its crew and drivers. It got a different paint job for Sebring, the fin's stripes were gone.

Mitchell's SR-2 suffered various mechanical problems during the 12 hour race. Track officials are in striped hats.

Still, the SR-2 was competitive.

And it finished the race, placing 16th.

(Left) Illuminated by a flash, spectators examine Mitchell's SR-2 after the Sebring race.

After the Sebring race, the headlight and parking light fairings were removed. These photos are dated April 26, 1957.

The Harley Earl SR-2 with its new, larger, headrest/fin in place. The person is not identified; could he be Jerry Earl?

Jerry Earl's number one SR-2 running at Road America in June 1957, with Curtis Turner driving. (Jerry McDermott photo)

Rear and front views of the Earl SR-2 at Elkhart Lake, June 1957.

Jim Jeffords driving the Purple People Eater in 1958 for Nickey Chevrolet. (National Automobile Historical Collection, Detroit Public Library)

Current owner of the number one SR-2, Rich Mason, driving the car at the Historic Sports Car Races at Laguna Seca, Monterey in 1990. (Ron Burda photo)

The third SR-2, made for GM President Harlow "Red" Curtice, poses on the east side of the Styling Studios.

Corvette Super Sport

Chapter 5

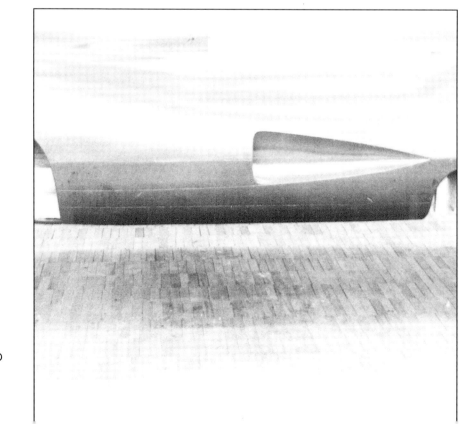

The future of the Corvette was in doubt shortly after its introduction to the public in 1953. The car gained very little acceptance by — and following within — the general public up through 1955. Chevrolet's engineer Zora Arkus-Duntov knew the Corvette needed improvement, and it needed the public's attention.

In 1954 Zora Arkus-Duntov presented a paper to the annual convention of the Society of Automotive Engineers (SAE). The paper was, in effect, Zora's racing rationale. It explained what sports cars were, why they needed higher performance than family sedans, and why they had to be raced in order to attract performance-minded enthusiasts as buyers. "I remember that the thrust of my talk was to convince the audience that the performance of the sports car must always be superior to the passenger car," Zora recalled, not sure he convinced anyone. "Of course, in the case of the Corvette, this was not yet the case."

"Then I said another thing. Race the car, not the car you sell to the public, but a special racing version. The Jaguar had the D-type for racing, and other models for public consumption. The Porsche had Spyders for racing, and others for sale to the public. This is what I felt Chevrolet must do with the Corvette."

In 1956 Corvettes got publicity and recognition because of: 1) Duntov's high speed run in Daytona Beach at 150 plus mph. 2) Corvettes successes at the Daytona Beach Speedweeks time trials. 3) Corvettes racing at Sebring. And, 4) Corvette's performances in SCCA racing, where Dr. Dick Thompson drove Corvettes to a national championship in the C production class.

The publicity from these events gave Corvettes much needed attention. In addition, Duntov and the production engineers were improving the 1956 production Corvette. Potential purchasers visited their Chevrolet dealer, and sales were increasing. For the first time, the future of the Corvette looked promising.

Duntov finally got his special Corvette, called the Corvette Super Sport. It was commonly known as the Corvette SS, or sometimes just SS. There was one Corvette SS test vehicle, called a mule. Although three Corvette Super Sports were planned, only one was started in late 1956 and completed in 1957. The Corvette SS was scheduled to be raced at Sebring and Le Mans in 1957. The story of the planning, development, construction, and racing history of the Corvette SS are detailed in this chapter.

There was a third Corvette Super Sport. This was a stock Corvette modified into a special car for the auto show circuit. Because it had the same name as the racing version, there is some confusion. The third auto show version of the Corvette SS, is discussed in detail in the Auto Shows section in this book.

The history of the 1957 Corvette SS begins with Duntov's 1954 SAE paper. Yet by spring of 1956, in spite of Corvette's racing success and Duntov's urgings, the special racing Corvette was not being developed.

Harley Earl, the head of GM's styling section, hadn't been directly involved with Corvettes since he had the idea for the original Motorama car. It was the unexpected influences from Jaguar and Mercedes that started the process. Earl borrowed a D-type Jaguar that had raced at Sebring in 1956 and placed third. Sometime after the race the engine was over-revved and destroyed. Earl wanted a racing Corvette, so he planned to install a Corvette engine, in the D-type Jaguar

CONFIDENTIAL

GENERAL MOTORS TECHNICAL CENTER

INTER-ORGANIZATION LETTERS ONLY

DATE September 11, 1956

SUBJECT Minutes of Car Design Advisory Committee Meeting on the XP-64

TO (All those attending)　　　　ADDRESS

Attended by: Messrs.　W. L. Mitchell　　Messrs.　J. Gilson
　　　　　　　　　　　J. Andrade　　　　　　　　　W. Hess
　　　　　　　　　　　R. J. Lauer　　　　　　　　　J. Himka
　　　　　　　　　　　C. M. MacKichan　　　　　　R. Cumberford
　　　　　　　　　　　R. Veryzer　　　　　　　　　D. Probst
　　　　　　　　　　　R. McLean　　　　　　　　　W. Dennis
　　　　　　　　　　　C. C. Whittlesey　　　　　　H. Elwert
　　　　　　　　　　　R. Potocnik　　　　　　　　K. A. Pickering

Meetings of the above personnel were held at 3:00 p.m., September 10 and at 3:00 p.m., September 11, to discuss the design, engineering, and fabrication of the XP-64.

The following decisions were reached:

OBJECTIVES

The XP-64 will be a competition racing car with special frame, suspension, engine, drive train, and body.

Since Harley Earl dreamed of developing a special racing Corvette, he had his styling studios prepare several drawings and renderings in July of 1956. Early drawings of the proposed racing car used the Corvette toothy grille and hood bumps on a low, streamlined body with a Cadillac-like fin.

This is the D-type Jaguar, with its empty engine compartment, which Harley Earl borrowed to push upper management into authorizing the special racing Corvette.

body, and enter it in Sebring as an experimental Corvette.

In June the D-type Jaguar arrived at Studio Z in good condition, but, without an engine. From the book *Corvette, America's Star Spangled Sports Car*, Bob Cumberford, recalled: "We were supposed to figure out how to install a Corvette engine, convert the steering to left hand drive, and disguise the body so no one would ever guess that it was a Jaguar. All without changing the aerodynamic qualities!" The appropriate plans and sketches for this transformation were prepared.

When Zora Arkus-Duntov learned about the hybrid, he expressed immediate reservations. There were many serious drawbacks to the major modifications required. Duntov proceeded to sketch out a new chassis for the racing Corvette he had described briefly in his 1954 SAE paper. He wrote a proposal for the design and construction of a special racing Corvette, which was approved.

The borrowed D-type Jaguar was returned to its owner unchanged. The question about intent remains. Was Harley Earl serious about modifying the D-type Jaguar to produce an experimental Corvette? Harley Earl was known for his ability to push and make things happen. Was this just a bluff by Harley Earl? We may never know, but the Corvette SS program wouldn't have existed without Harley Earl's influence.

The project to build a special racing Corvette was approved on August 7, 1956. It was assigned project number XP-64. This would be the official code name for the project. XP-64 was also assigned S.O. 90136, which authorized a full sized clay model, called a clay buck. The shop order (S.O. 90136) was a way of keeping track of labor and material expenditures on approved projects. The clay buck was to be completed on August 15.

Harley Earl's styling studios had already been turning out drawings. But this was not the final version: drawings were completed, then revised and redrawn. The final version of the XP-64 changed several times to suit Styling's upper management, which included Mr. Earl and Bill Mitchell. Its contours were influenced by the D-type Jaguar, a much-admired car among the stylists.

A September 4th memo reads: "On Friday, August 31, Mr. Earl outlined the forthcoming program on the XP-64 lightweight high performance car. This car is being modeled in the Chevrolet Studio." Also, "Mr. Duntov of Chevrolet Engineering will represent the Division on this project."

With Harley Earl in charge he appointed the following committee to administer the program: R.F. McLean, C.C. Whittlesay, J.S. McDaniel, and Del Probst. The memo spells out their duties then continues: "Mr. Earl has directed that three identical vehicles be produced and ready for delivery by December 15, 1956."

Now, this is the first time that the time frame had been mentioned. The Sebring race was scheduled for March 23, 1957. If the three identical XP-64s were ready by December 15, 1956 that would be plenty of time to get them ready for the race.

The next memo is dated September 11, 1956, and states in part: "Chevrolet and Styling Management reviewed the XP-64 clay model in the Chevrolet studio September 10, 1956. It was decided at this meeting that the XP-64 would be shown in New York at the Automobile Show December 8-16." Also a list of nine items to be modified was included.

The racing Corvette was authorized on August 7, 1956. The final design was determined, and a full sized clay model was nearing completion after only two weeks.

The meeting held September 11, 1956 resulted in several important statements:

"OBJECTIVES: The XP-64 will be a competition racing car with special frame, suspension, engine, drive train, and body.

PRELIMINARY PROGRAM PLANNING, STAGE NUMBER ONE:

"Tentative completion of clay model - September 14.

"Seating buck and instrument panel buck will be started immediately in the Experimental Interior Design Studio.

"Completion of the XP-64 automobile was scheduled for Saturday, December 1, for a showing in New York on December 8.

"This will be a "glamour" automobile with magnesium body and as realistic in appearance as possible." In other words, folks, this was going to be a non-running show car.

STAGE NUMBER TWO: Completion of three additional XP-64 automobiles at some later date, to permit testing prior to the Sebring race in March.

"These racing cars will be complete competition automobiles in all respects, including magnesium bodies and titanium exhaust systems."

Still another memo dated September 11, 1956 was sent to Mr. W.L. (Bill) Mitchell: "This will confirm the instructions received this date, from Mr. E.N. Cole of Chevrolet Division, to the effect that the XP-64 will not be shown in New York City on December 8.

"However, three versions of the XP-64 will be completed about the first of the year, for testing prior to use in March."

Well, it seems Chevrolet General Manager Ed Cole was to have the last word on the XP-64 being shown in New York. In addition, the completion dates for the three cars has slipped back about two weeks.

September 12, 1956 memo, "Authorization has been given to construct a preliminary seating buck for the XP-64 clay model, S.O. 90143. This buck is to be completed with all possible haste for a showing to the Management at 8:30 a.m., Monday, September 17, 1956."

September 14, 1956 memo, "a meeting was held in the auditorium at 10:30 a.m. Friday, September 14, to evaluate the XP-64 clay model in comparison with the D-type Jaguar, Ferrari, and 1956 production Corvette." The memo included a list of eight changes to be made to the XP-64 clay model, plus this: "Information is urgently needed on the following items to complete the clay model:

1. Gas filler location and function.

2. Steering wheel location.

3. Revised wheel geometry as a result of increasing the rear tread from 49" to 51-1/2"."

September 17, 1956 memo, "SUBJECT: Minutes of Meeting on XP-64 Seating Buck, S.O. 90143.

"After a preliminary examination, it was determined that none of the components in the proposed locations could be used. With Mr. Duntov standing in as "driver," the various components were relocated in positions approved by those present. "In all, fourteen changes were called out like clearances, pedal re-locations, etc."

September 21, 1956, a meeting was held to discuss problems on the XP-64. Nine problems and their solutions were discussed.

The clay modelers shape the SS body, while technicians document the shape on a drawing in the background.

Duntov had planned on a tubular space frame, but the car was due to race at Sebring in less than six months. With little time for development, Duntov used a borrowed frame of a Mercedes-Benz 300SL to be used as a pattern. The SS frame was fabricated from round or flat tubing, the final frame being quite different from its 300SL mentor. The chassis weighed an acceptable 180 pounds.

By October 1, 1956 the final dimensions of the XP-64's wheelbase had not been determined. It seems the back of the seat was too close to the rear wheelhouse. The possibilities are 1) moving the passenger compartment forward or 2) extending the wheelbase. A decision will be made later.

Another October 1, 1956 subject was lighting. Since the car would be racing in the dark part of the time, consideration was given to angled lights for the sides of the race course. From the memo:

"LIGHTING: Mr. John Yee presented a set-up to compare the various lights under consideration. These were:

1. 7" production guide T-3.

2. Dual 5-3/4 production guide T-3.

3. 7" Lucas Le Mans driving light.

4. Two 5" Lucas driving lights.

5. One 5" Lucas high beam-low beam driving light.

6. The Cibie lamp was not available at this time.

"These lamps were compared with regard to intensity and light pattern. All those attending were of the opinion that such a static evaluation is of no value, that these lamps should be mounted on a roadable car for final testing. Mr. Yee will make the necessary arrangements." Also, "Mr. Duntov and Mr. Lauer both mentioned that the deadline date for the completion of the roadable car has been established for January 15, 1957, and all design, engineering, and fabrication activities should be geared to this date." The completion date just slipped back another two weeks.

October 3, 1956, another meeting was held to settle the seating-wheelbase problem. No final decision was made.

October 4, 1956, the wheelbase varied from 90.5" to 96", as proposed by various styling and engineering groups. It was finally set at 92" on this date.

Meanwhile, Duntov was working on the frame and chassis details. The front suspension was to be fabricated from sheet steel. An independent rear suspension was considered, but it was an unproven system. Duntov preferred the de Dion rear axle, because he had worked with it before. The de Dion system used a steel one piece tubular axle which curved around behind the quick-change section of the Halibrand differential.

In order to test the de Dion axle and determine its best location, Duntov used a conventional Corvette chassis. The front of the chassis, including the frame, engine, transmission, and front suspension were retained. Its entire rear frame and suspension system was removed and the de Dion rear axle was installed. This is how the de Dion system was tested, this also helped establish the wheelbase dimension for the XP-64.

Under the approval for XP-64, one test vehicle and three competition cars were to be built. The three cars were to be ready for Sebring 1957, and were also to be raced at Le Mans, France. With the wheelbase decided, Duntov proceeded with the test car frame, which picked up the standard test car nickname, mule.

September 1956, developing the design of the Corvette SS in clay. The GM design staff are, left to right, Frank Funk and John O'Brien, clay modelers; Bob Veryzer, Assistant Chief Designer, Chevrolet Studio, and Jack Park, chief clay modeler.

The specifications for the three XP-64 cars called for a body to be fabricated from sheet magnesium. This is a bit surprising, because GM had been a pioneer in the fabrication of fiberglass car bodies. The mule test car, however, was to have a fiberglass body. It was expected to become damaged and modified, so it was to be painted simply "Alpine White inside and out."

October 11, 1956, the minutes of the XP-64 meeting, "1) Mr. Gilson will engineer a chassis mock-up to be covered with as-is-fiberglass skins, plastic instrument panel, and dummy underpan. This buck will be used for engineering placement and clearance studies plus other problems that may be better solved three-dimensionally. It will also be used as a seating buck, and will contain seat, pedals, and tunnel.

2) Mr. Gilson will engineer a full size wind tunnel test model, utilizing fiberglass body skins, sanded to a smooth finish, and a metal underpan. This model will be used to study air flow around the body and the value of the interior ducting required. The model must be fully reinforced and internally ducted, probably with cardboard and tape, to simulate actual conditions.

3) Mr. Hess will utilize the present seating buck with an up-to-date cowl and cockpit opening for the instrument panel clay modeling.

4) A mandrel will be constructed from reinforced fiberglass skins to aid in the forming and the fitting of the actual magnesium skins.

The following dates were tentatively established:

    Completion of Clay Modeling .... October 17

    Completion of Recording Engineering Information from Clay Model ...... October 23

    Completion of Fiberglass Mandrel ...... October 30

    Completion of Wind Tunnel Test Model ...... November 9"

Using the full sized clay body developed by the styling studio, a wood form and a plaster mold were built. Two fiberglass bodies were fabricated in the mold. After the white paint job, one was installed on the mule's chassis.

The other fiberglass body was sanded smooth and also painted white. It was built up to look like a real vehicle, although it had no chassis. Tufts of wool were placed all over the body, and it eventually went through an air flow test in the Research Dept.'s wind tunnel. The results were compared to the D-type Jaguar, and it was reasonably close, so it was accepted with no modifications.

Zora Arkus-Duntov was authorized to build a test chassis. He liked the de Dion rear suspension system, so a de Dion was installed on the rear of a stock frame. The instruments were mis-matched; the body was odd-shaped panels screwed together; the exhaust system was a muffler hanging on the side. Yet this crude chassis guided the engineers as to where and how to install the de Dion rear end.

Using a Mercedes 300SL frame as a guide, Duntov designed a tube frame for the special Corvette test chassis.

A technician begins to install the de Dion rear suspension system on the tubular frame.

The front suspension and heavy duty front brakes are installed.

The wheels, engine, transmission, and fuel tank are added to the test chassis, which is now called, "the mule."

A close-up of the engine in the test mule. Note the 220 hp base engine with stock exhaust manifolds present, as this is a chassis development car. Also, note the use of plywood in the firewall and other panels.

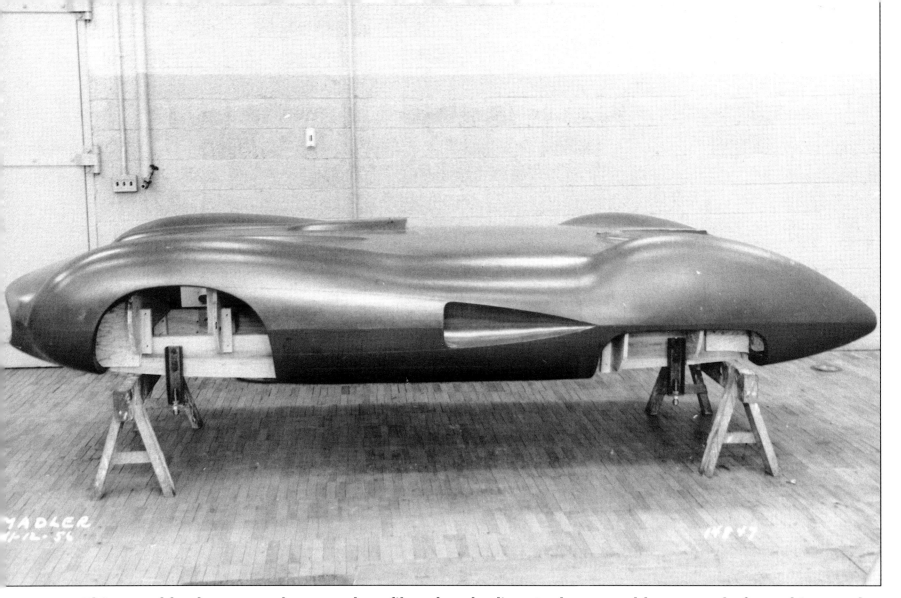

This wood buck was made to produce fiberglass bodies. A plaster mold was made from this wood buck. Using the plaster as a female mold, two bodies were built using the hand lay-up method. One body was made for the mule, the other was installed on a dummy chassis and used in a wind tunnel test.

The fully functional fiberglass body is installed on the mule chassis.

The wood grille from the clay model is installed, and the mule is readied for test runs.

Viewed from the rear, the mule has no trunk: the fuel tank is mounted directly behind the driver, above the differential and the inboard brakes.

(Left) Duntov takes an automotive writer for a ride in the mule on the test track at the GM Tech Center in Warren, Michigan.

The styling studios were working on the internal details of the racing Corvette. This is the instrument panel mock-up, with wooden stringers showing the outline of the upper front body. Note the drawings and blueprints on the table in the background.

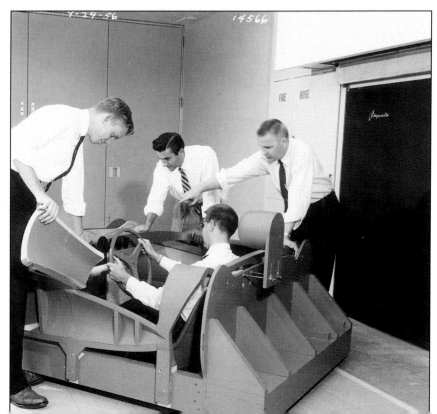

This is the seating buck, used to get the right seating placement.

CORVETTE: AMERICAN LEGEND

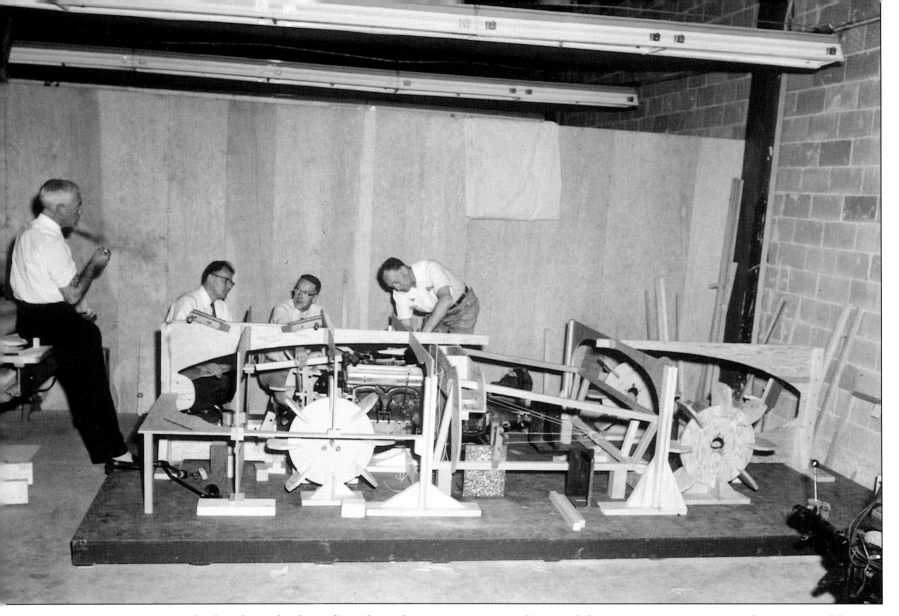

This is the body buck, which is fitted with an engine and wood forms to represent other parts like tires. The actual dimensions of the body are calculated, while carefully following the design of the full sized clay model.

The body buck is in the background, while the body dimensions are transferred to blueprints in the foreground. Note the center section of the quick-change rear differential on the table. Detailed blueprints were made for each section of the car, because three identical racing cars were to be built at this time. The full sized drawings had to be revised or redrawn every time a change was made to the original design. This attention to documented details took lots of time.

SS Construction

November 12, 1956, from the minutes of the XP-64 meeting, discuss the status of the project. "Material requirements for the magnesium body should include:

a. The original roadable car.

b. A spare set of body panels.

c. Possibly one or two other roadable cars to be fabricated at some undetermined date."

Note that, without a clear decision from management, the intent is now to produce one "roadable" car.

A meeting was called for November 9, 1956, to discuss the instrument panel layout. Included was this statement: "Since engineering is 60 percent completed and fabrication is 40 percent completed, there is no time to lose."

November 16 meeting, status of the XP-64:

MATERIALS

"Because of supplier difficulties, the desirable body material will not be available until January 15. As this date is far beyond the present schedule, it is planned to use the material now on hand, which is in sufficient quantity to build one car and a complete set of spare panels."

This is the reason that only one XP-64 was built. In the beginning, according to plan, there was one non-running car to be shown at the New York auto show, and three competition cars. The show car was dropped, leaving the three cars for Sebring. Finally, a shortage of sheet magnesium limited the project to a single vehicle.

The meeting turned to the forthcoming wind tunnel tests. Six items were to be completed on the model, including: "the wooden grille insert from the clay model." The wind tunnel test was scheduled for "sometime about the first of December."

Also discussed at the November 16 meeting, "HEADLIGHTS, on the basis of actual road lighting tests conducted thus far, all indications point to the use of the Cibie 7" headlight. Because of the delay in obtaining such lights, orders for reserve parts should be placed in the near future."

There were several reports that Duntov had built the test car "mule" without authorization, and was done on the sly. But, that was not the case. Ed Cole was quoted in an article from the *Northern Automotive Journal*, "On October 1, 1956," Cole said, "Arkus-Duntov was assigned an engineering staff which began literally to build a car within reaching distance of the drafting boards. A rough prototype was first tested, then the original design was modified and construction began."

Another factor that slowed design of the XP-64 was the requirement to work from blueprints. It took time to document everything and revise the drawings. This does make sense, for the plans were to eventually build at least three cars for competition, and they wanted them all to be identical.

Duntov liked disc brakes, but they were in their infancy in 1956. Some units were available for light weight cars, but none that would work for the XP-64. From an article in *Speed Age*, "The brakes, using Bendix new 'Cerametallix' (bronze-ceramic) lining, seem to be adequate for the roughest going. The basic brake units, incidentally, are stock Chrysler Corporation 'Center-Plane' types — by far the finest passenger-car brakes in America today."

From *Corvette, America's Star Spangled Sports Car*, "They (Chevrolet's engineers) knew they could get good braking results from the Cerametallix lining used at Sebring in 1956. But, they wanted a two-leading-shoe brake mechanism, one that would give more consistent response than the usual American duo-servo brake. The necessary components: pressed steel shoes, pivots, retractors and wheel cylinders were all available off the shelf, but, unfortunately, not a GM shelf. They were 1956 Chrysler Center-Plane front-wheel brakes, used on all four wheels of the XP-64. This set the brake size at a 12-inch diameter and a 2-1/2 inch width.

On November 26, a meeting was held to review the status of the XP-64. Among the items discussed: " 'CHASSIS' styling fabrication body work will be completed to the point where a chassis is required on about December 20. Chevrolet indicates that the chassis will not be available at this time. Therefore, a mock-up will be constructed holding chassis attachment points for use in preliminary body assemblies. Body draft and skin information 90 percent complete." And this: "Overall exterior engineering is 40 percent complete and is ahead of Chevrolet Divisional work." One more item: "Wind tunnel model 95 percent complete. Requires only paint."

The next status meeting was December 19, several items have not been completed; including, the front suspension and the gas tank installation. From the memo, "Mr. MacKichan indicated that no promised date has been received on the delivery of the chassis. Styling is progressing with the chassis mock-up in order to expedite body assembly."

The XP-64 chassis had two unique systems, brakes and the ducted radiator. The brakes were power-assisted by two vacuum boosters. The vacuum booster located on the right rear corner of the chassis operated the front brakes. The booster located at the left rear corner of the chassis operated the rear brakes. A valve was located in the vacuum line to the left rear brake booster. This valve was actuated by a dash mounted mercury switch. Under hard braking, the mercury switch

The SS suspension parts are redied for installation on the chassis. The battery is a "Rebat" aircraft battery. Note the rolls of blueprints and drawings in the right rear background.

closed the valve, thus sealing the amount of boost supplied to the rear brakes. With the rear brake thus locked, the driver was able to apply hard braking to the front brakes: when the brakes were released, the system reset itself. It was a clever way to keep the rear brakes from locking up.

The radiator ducting also deserves a closer look. Air was ducted through the grille and radiator, then ducted up and out through louvers in the body at a low pressure point. This system delivered a downward force on the front of the car, helping to keep the front end from lifting.

"WIND TUNNEL MODEL: Mr. McLean indicated that the styling wind tunnel model has been entirely completed, including all static pressure probes and woolen tufts. This model will be shipped to the research wind tunnel, pending final arrangements with Research by Chevrolet Division."

In a memo dated December 20, "Mr. Premo replied as follows:

1) The wind tunnel test will be conducted December 27 and 28, 1956.

2) The target date for chassis completion is January 22, 1957."

From a January 7, 1957 memo, parts are being installed on the chassis. A scale has been provided, and all parts are to be inspected and weighed, and proper records kept.

From a January 16 memo, for the first time a note is hand written on the memo, "Sebring Corvette." Up until now the project was only known as XP-64. On the status of XP-64 bodies, "Number one body parts 80 percent complete." From the same memo, "Assembly of all completed parts will be started in ten days."

In reference to the spare body panels, "Panels and parts for Number two body have just begun." Then, this comment, "Number two body parts ready for assembly in four weeks."

On January 21, a major memo was released. It reported that the latest Sebring regulations had been received on January 17. Several meetings followed, as "the XP-64 did not conform to specific rules."

Briefly, the rules that required changes in the XP-64:

"1) All vehicles shall be equipped with at least one rigid door on either side, with a closing device and hinges, giving direct access to the seats.

2) A hood (top) shall be required and shall have to be fixed on the car for presentation to preliminary inspection.

3) Under no condition may the seats serve as a holder to a spare wheel or be combined with the fuel tank.

4) The windshield is compulsory.

5) The windshield must have an automatic wiper."

After a listing of these problem areas, "It was decided that the model should be changed to comply with all regulations."

Suddenly the XP-64 required a removable hard top. This extra project was handled thus, "Chevrolet studio will begin design studies on a top for the subject model. This may be constructed from Plexiglas and have an integral upper windshield. It is understood that this top will be removed during the race."

Other details, "Mr. Gilson has instructed Trico to continue with engineering and fabrication work on a suitable set of windshield wipers for this model. It is not planned to use the wipers during the actual race."

The non-functional "dummy" engine is lowered into the SS chassis.

These changes took time, of course. The time slot was beginning to accumulate and delay the project. Now we have this statement, "A tentative delivery date of the chassis was established for approximately January 30."

On January 31, 1957, a memo was sent to Zora Arkus-Duntov from styling. It listed the items to be installed on the chassis before the body is fitted. It concluded with this paragraph, "At our meeting, you indicated that the chassis described above would be available approximately the fifteenth of February. Styling will make plans based on this date. We are planning this chassis, with the body mounted, will be the final unit for the competition of the car."

An internal memo from Styling dated February 11, 1957, listed seven items to be corrected on the body. The final completion date for everything was to be February 16, 1957."

A memo dated February 18, 1957, addressed the final assembly schedule. This was barely over one month before the race date of March 23. The schedule, "It was agreed that Chevrolet would deliver the bare chassis frame without components to Styling on Monday, February 18. Styling would complete the attachment of mounting brackets as part of this frame and proceed with all possible haste to fit and assemble a magnesium body. It was planned to complete this work by Monday, March 6th, and return the body and frame to Chevrolet for completion of chassis assembly, road test work, and etc. The car was then to be returned to Styling p.m. March 11th for painting, trimming, and final assembly before loading for truck delivery on Saturday, March 16th."

The memo continues, "However, at a subsequent meeting on Saturday with Mr. Premo, Chevrolet made it clear that this program would be unsuccessful unless the complete automobile can be delivered to Sebring, Florida, around the 5th of March for two weeks of high speed road test work on the proposed course." At this time it was pointed out that their plans include shipping by air of the completed automobile with body on Monday, March 4th.

There's more, "While it does not seem feasible for Styling to meet this request, it was agreed that everything would be done to come as close to this date as possible. To complete this body in paint and trim in the fourteen day schedule requested, it will be necessary that we take advantage of every available hour and that we compromise on finished workmanship and appearance to a great extent."

Now the schedule is very tight. All the engineers and technicians were working overtime to get the XP-64 car completed on time. The car is in Michigan with one month to go. The XP-64 must be completed and shipped to Sebring, Florida before the March 23rd race date. A memo dated February 20 indicated that there was still more work to do on the XP-64 body. One hold-up in the schedule was the quality of work demanded by the Styling upper management. They were producing a show car finish on a race car.

In a memo on February 27, Styling management stated, "I placed this program on a 24 hour, 7 day a week basis and advised them (Chevrolet) to wait until we could complete the body here, approximately March 5th, before (shipment) to Sebring."

In regards to the car while at Sebring, "The construction of this extremely light weight magnesium body requires that it be considerably dismantled for any real work on the chassis. We consider that first-class appearance, as well as the proper assembly of this car, is of paramount importance. It is doubtful whether the Chevrolet people would give this same concern to

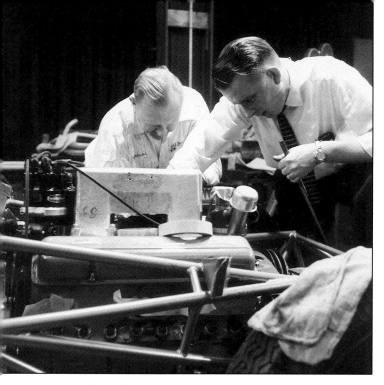

Technicians and engineers document every dimension and revision on full sized drawings, which took a lot of time.

The dummy engine is used to check clearances, so the wood fuel injection unit serves that purpose well.

Chet shows us the Chrysler based brakes, with a wheel cylinder for each brake shoe. The brake drum is an Alfin, a cast iron braking surface with aluminum fins cast directly to the cast iron. Duntov improved the Alfin drums by drilling a number of holes around the perimeter of the cast iron drum. When molten aluminum was added, it penetrated the holes and the result was better brakes because of improved heat transfer.

106 CORVETTE: AMERICAN LEGEND

appearance during the test period." The Styling folks are still concerned about appearance.

A memo dated February 28, 1957, gave details of the XP-64 lighting. Of the ten items listed, here are a couple: "WINDSHIELD WIPERS: Two individual constant up and down switches will actuate wipers individually or separately at full sped in down position. Resistors to be placed in wiring for half speed in up position. Switches mounted on instrument panel."

"TAIL LAMPS: There are three lenses on each side, the center lens of each set of three are reflectors only. The other four, which have double filament bulbs, are lighted simultaneously on the brighter filament for stop lights and simultaneously on the dim filament for parking lights. These lights to be wired to treadlight switch and brake light switch."

The last minute changes to the XP-64 as required by the rules at Sebring delayed the car several days. The show car appearance insisted on by Styling delayed the car even further. Following the tradition of the GM Motorama show cars, the car had its finishing touches applied in the truck on the way to Florida.

Meanwhile, at Sebring, several world class drivers had been hired to drive the Corvette XP-64. Juan Manuel Fangio had signed a contract to drive the XP-64 at Sebring. As race day approached with no race car in sight, Fangio was released from the contract. Chevrolet had wanted Stirling Moss to co-drive with Fangio, but Moss had other obligations, so a contract was signed with Carroll Shelby. Without the car, he too was freed from his contract.

Chevrolet had hired John Fitch to manage the team of stock and semi-modified SR-2 Corvettes. Fitch agreed to drive the XP-64, and he suggested veteran Piero Taruffi as his co-driver. Taruffi flew in from Italy at the last minute, and was able to get a few laps in the mule.

Duntov and his crew had the mule's brakes adjusted and broken in, and ready for the practice session on Friday. Lacking the XP-64 itself, the mule was driven many laps around the 5.2-mile course in practice. From Bernard Cahier's article in *Road & Track* about Friday, the day before the race, "Everyone was waiting to see one of the main attractions of the practice, the new Chevrolet Corvette in action. The one planned to run in the race as not quite ready, so it was a beat-up looking training car, which appeared on the track during practice. With Fitch at the wheel the car did 3:32 and Taruffi did 3:35, so everyone was impressed as well as surprised. Toward the end of practice, Fangio was invited to try out the prototype. With his typical grin, Fangio took off for the first time in this car. Within three laps the "Old Maestro," who seemed to be enjoying himself immensely as I watched him on a curve, had accomplished the fantastic time of 3:27.4. He was justly greeted by warm applause. The Americans could well cheer about that performance, since Fangio had proven to them that, if masterfully driven, the new Corvette could hold its own with the best from Europe. This was also good news for the sport, as any good newcomer, and especially an American one, has a most stimulating effect on good motor racing. Moss also tried the new American threat and turned a one lap performance in the Corvette. Especially praised was the road handling of the car, its brakes and its comfort. Fangio even told me that it seemed that he could drive this car comfortably at high speed for 10 days if he had to!"

Another description of the SS test car is found in *Corvette, America's Star Spangled Sports Car*: "Actually, to call that test SS an automobile was a form of flattery. Its interior was a mass of tubes and mis-matched instruments; its body was

The rear suspension parts: the technician is holding a stabilizer link whose rubber bushing failed during the Sebring races.

Installation of the front brakes.

Close-up of the rear shock and its exterior spring.

rough fiberglass with no doors and no rear-deck cover, mounted over a one-inch plywood firewall. The mule, as it was known, was 150 pounds heavier then the race car and was down on horsepower, but it had a full quota of Chevrolet/Duntov engineering."

The XP-64 became simply the Corvette SS when it was delivered to the Sebring track in Florida, the day before the race, March 22. Much of that day was taken up in publicity: photos of the car with different drivers in various poses. Finally, Fitch got to drive the car a few laps. The brakes locked up, which was characteristic of this lining. Until the brake lining got broken in and heated up, the car swerved in one corner, then again in the next corner as the brakes locked up.

Heat was to be a big problem with the Corvette SS. From *Corvette, America's Star Spangled Sports Car*: "Even though the race SS was never really pushed in practice, one feature of its design was immediately evident, it was sizzling hot in the cockpit, 'intolerable', to quote Duntov. Its magnesium body conducted heat, while the fiberglass body of the mule had insulated it, and the rough-built mule was far better ventilated. Mechanics rushed to cut away the lower body panels along the exhausts; insulation was packed inside the cockpit, and scoops and ducts were added to the doors on both sides. It was still unbearably hot in the car."

Although about two thousand miles had been put on the mule, the Corvette SS itself was virtually untried when it started the race at 10:00 a.m. on March 23, 1957. Perhaps it was worse than an untried car. Fitch had only a few laps driving the car. The night before the race, after practice, new brake linings had been installed in an attempt to eliminate the locking brake problem. It was assumed that this would solve the brake problems.

Fitch was able to take the car out for a few miles on an unused section of the Sebring airport Saturday morning just before the race started. The brakes still locked up, but there was no more time. With 15 minutes to go, the SS was pushed to the starting line. Alas, the new brakes locked up from the beginning. After two laps, a flat spot was worn on the right front tire, causing high-speed vibrations. On the third lap Fitch came in for replacement of both front tires.

Installation of the rear brakes which mount directly on the quick-change rear differential.

The rear differential with its inboard brakes installed in the SS chassis.

Several engineers and technicians at work on the chassis.

CORVETTE: AMERICAN LEGEND

The all-aluminum T-10 four-speed transmission is inspected and readied.

The all-aluminum four-speed transmission is installed in the chassis.

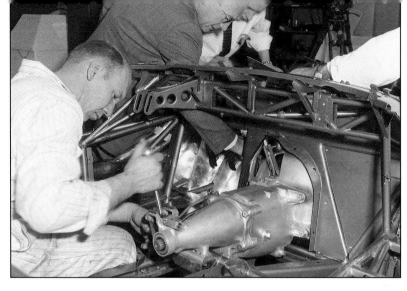

Final testing after installing.

An overview of the chassis with the transmission installed.

The exhaust headers are added to the dummy engine: clearances between parts will be watched closely now.

Duntov and project foreman Ed Donaldson review the all-important drawings as the radiator is installed on the SS.

(Right) A closer look at the SS shows the radiator inlet and outlet ducts, designed to provide a downward force to the front of the car.

While the SS chassis is being designed and assembled, the full sized wood buck has produced a plaster mold, which was used to make two fiberglass bodies. One was mounted on the SS mule chassis. The other body was mounted on a non-functional chassis for wind tunnel tests, and is shown here compared to a stock 1956 Corvette.

CORVETTE: AMERICAN LEGEND

The body on the non-operational chassis was sanded smooth and painted white for the wind tunnel tests. It is shown here with four-inch tufts of wool attached all over the body.

The wind tunnel model shows the air flow patterns.

The base of the instrument panel is installed. Note the vacuum cylinders for the brakes on the rear corners of the chassis, with the battery tray in between.

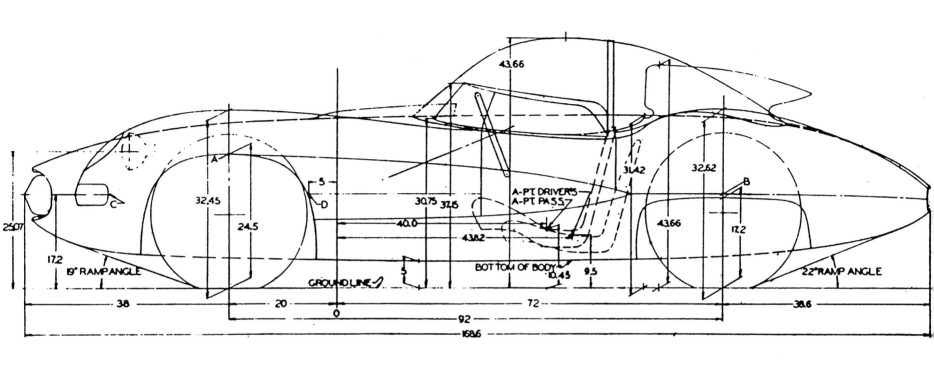

The design and dimensions of the Corvette SS are finalized.

Because the SS will be racing in the dark at high speed, good lighting is essential. Several types of lights might be used for the angled corner lights; here a stock Corvette is tested with various lights. Note the unusual side trim in the cove.

With the dimensions finalized and documented on the drawings, the all magnesium body is hand formed and ready for painting.

Meanwhile, the special fuel tank arrives and is installed.

The chassis is finished, and being prepared for the "real" engine.

The operational SS engine just before installation in the chassis.

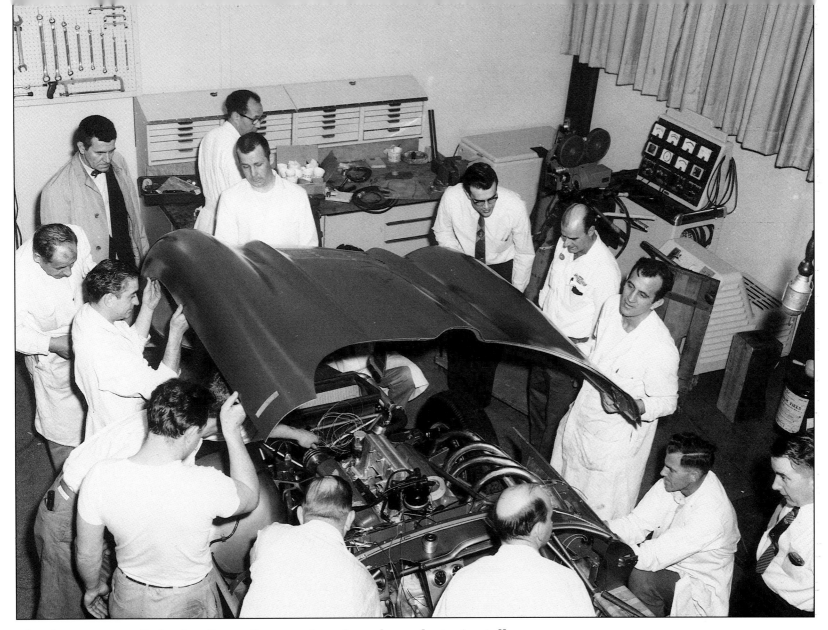

The front part of the body arrives painted and ready to install.

The body is fastened to the chassis.

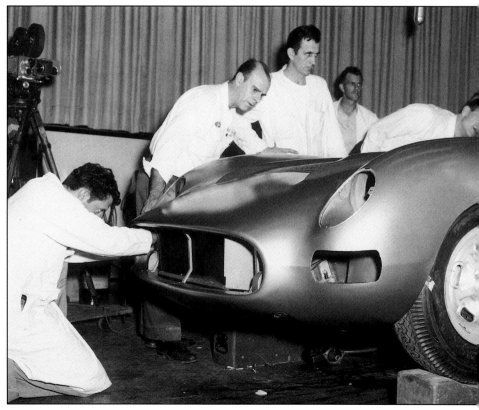

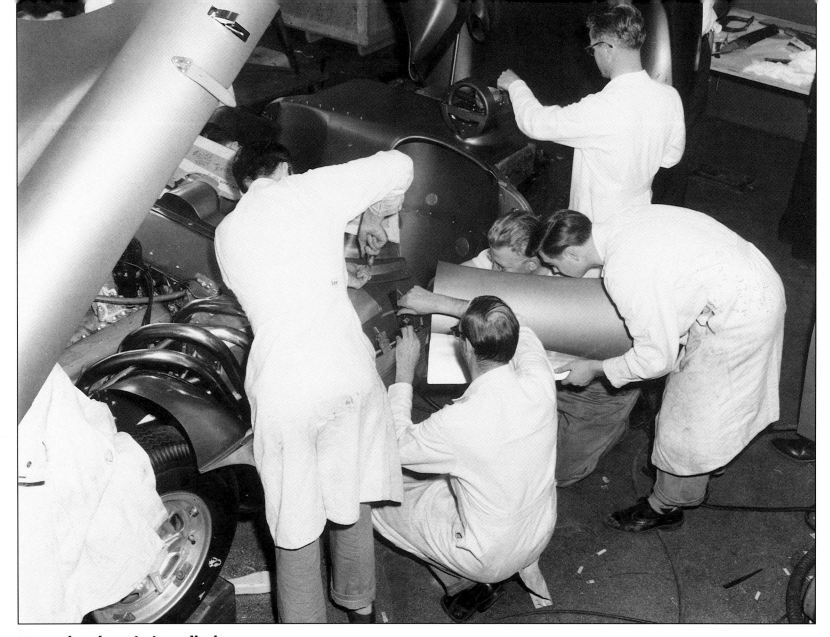

Next the door is installed.

The grille and the front and rear "CORVETTE" decals are installed.

A view of the engine and its magnesium oil pan from below.

The bubble top, a last minute addition, is added to the SS body.

The SS body is touched up again: it must have a show car finish, which delays the project many times.

The Corvette SS with the bubble top installed is shown in the Tech Center viewing yard for the approval of upper management.

As always, Chevrolet Engineer Zora Arkus-Duntov was nearby; here he sits in the SS.

Instrument panel of the completed SS. Instead of a telescoping steering column, three different steering wheels with varying offsets were supplied.

(Right) At last, the completed SS with both ends open. Note how the front of the headrest tilts forward, exposing the built-in roll bar and gas tank filler neck and cap.

The engineers and technicians roll the SS out for its weigh-in.

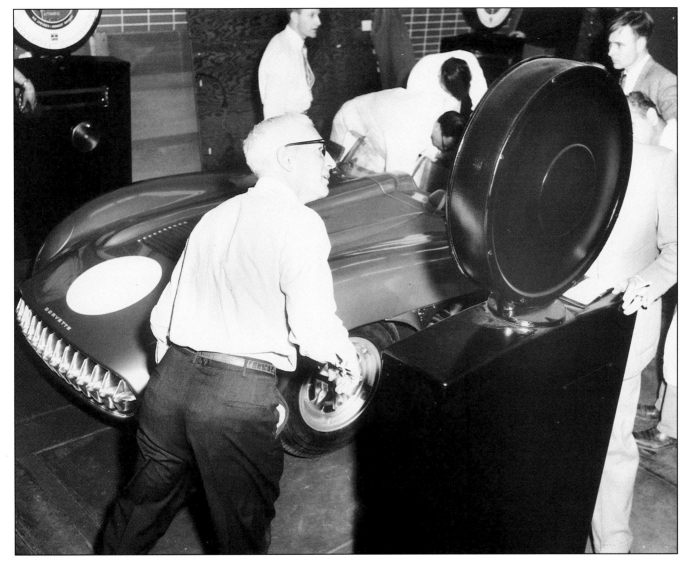

The SS weighs in within design limits; Duntov checks the scale.

SS Track Performance

From the book *Corvette, America's Star Spangled Sports Car*, "Pressing on, in spite of brakes that were still not predictable, the SS began to reduce its lap times, clocking 3:32.8 more than once and cutting one at 3:29.8, a clear indication that the race SS was an even quicker car than the mule prototype. Out of the corners it could pull away from the D-type Jaguars. Acceleration was already better than the 3.5 Ferraris and about par with the 3.8's, but it couldn't match the violent pull of the 4.5 liter Maserati that went on to win."

Suddenly the engine died completely, but it happened close to the pits. Fitch coasted into the pits with a quiet engine. There had been several problems with the fuel injection system, so no one suspected the ignition system. Sixteen minutes were lost until someone jiggled the coil wire and the engine started. Out on the track, the driver was permitted to work on the car with the parts and tools he carried in the car. So Fitch threw a coil and tools in the car and was off again. Soon the coil failed again, and Fitch replaced the coil out on the track. That eliminated the problems, and the SS engine ran smoothly for two hours after that. At this time the SS was about 20 minutes behind the race leaders, but it was turning good lap times, about 3.32 to 3.35.

Suddenly, the Corvette SS developed handling problems and was withdrawn from the race. From an article in *Speed Age*, "Just to keep the record straight, what did happen to the car in the Sebring race? Factory public relations men passed it off with the bland statement that they had learned all they needed in three hours, and were withdrawing from competition! Trackside rumors and press reports got all fouled up because of the mistaken idea that there were two SS models running in the race." The other car was probably the SR-2, which wore Number Two.

The Corvette SS mule was driven many miles on GM tracks in Michigan. Its handling and brakes were adjusted for optimum performance. It is shown here just before shipment to the Sebring track in Florida.

The engine in the SS mule has been changed and updated. Perhaps it's the same engine block, but now it has a fuel injection unit and headers. It was necessary to upgrade the mule, for the Corvette SS itself was late arriving at the Sebring track. (Ludvigsen Library)

From the Roger Huntington article in *Speed Age*, "What finally put the car out of the race was a parts failure in the de Dion rear suspension. The rubber bushings for the control links on the left side deteriorated sometime around the third hour. Since these were rather large bushings (actually a Plymouth part), this allowed over 1/4-inch play of the wheel on one side. This, in turn, caused the "steer" characteristics of the whole car to wobble rather violently and unpredictably between oversteer and understeer, both on the straights at speed and in the corners. Apparently the bushings were just starting to go when Fitch came in to refuel and hand the car over to Taruffi. Taruffi brought it in after one lap and said the car didn't handle right. Fitch took it around again, and it was obvious that something was radically wrong. It was decided to withdraw at this point, as the car was neither safe for the drivers nor the other cars on the course. The bad bushings were discovered later."

Another view of the problem, from *Corvette, America's Star Spangled Sports Car*: "... the rear suspension began to feel distinctly odd, to the extent of letting the tires touch the body and chattering uncontrollably after bumps. Undriveable and overheated, the Corvette SS retired officially after its twenty-third lap..."

And more from the same book, "The main cause of the car's retirement, the handling deterioration, was traced to failure of a rubber bushing at the chassis end of one of the lower rods that provided the de Dion tube with lateral location." As Duntov later said, "It was doomed to fail." If the fitter was not familiar with the installation procedure for these bushings, he could split them during the assembly of the joint. This one had been split and thus became a built-in focal point for failure. The design itself had not been to blame.

After the race, Duntov stayed in Florida for a well-earned vacation. While still in Florida, he was called with the news that the Automobile Manufacturers Association (AMA), was considering a ban on racing. This ban had been proposed by GM President Harlow "Red" Curtice, and others, at a meeting in February.

Even with the future of the Corvette SS in doubt, the car was still being prepared for the Le Mans race in June. On April 2, a list of 28 modifications was to be made, along with eight observations on the best ways to proceed.

The mule's body was removed and the mule's chassis was prepared for a new magnesium body. But this project, to build a second Corvette SS, was never completed.

Improvements on the SS continued through April. For example, take April 4, when new door hinges were being designed and cast. The entire project came to an end when an order came down from management in early May. The SS was allowed to survive, but all other components were to be scrapped.

The next memo came on May 29, when the SS was to be restored to show condition. Its racing days were over. The AMA ban became official on June 6, 1957. The SS was retained strictly as a show car.

Evidence of the Corvette SS as a show car was contained in a memo dated June 12, 1957. The Chevrolet Sales Section had permission to exhibit the car at the Elkhart Lake event next week.

And this statement, "At their request (Chevrolet Sales), this car will be repainted in a plain blue color without white circles or numerals. Also, any designation of this as a "SS" racing

Thursday, March 21, 1957, the first official practice day. Several different drivers have driven the mule. The thick fiberglass body and some iron components (like the transmission case) have the mule weighing about 150 pounds more than the magnesium-bodied SS. Here Duntov has just driven the SS (see the sign), and is comparing notes alongside the mule. (Ludvigsen Library)

The SS mule is looking a bit worn and ragged, but it's still capable of turning fast laps. (Bernard Cahier)

vehicle will be eliminated as they are considering it as merely an experimental Corvette." How the mighty have fallen!

The Corvette SS was used again as a test vehicle by Zora Arkus-Duntov in December of 1958. Duntov was turning 183 plus mph at the test track at the GM Proving Grounds in Arizona. When the new Daytona track opened in February of 1959, Duntov drove the SS at 155 plus mph.

Then, on May 15, 1959, "Chevrolet Division has notified Mr. MacKichan that the subject XP-64 (SS) Corvette will not be required for display purposes at this time, and therefore, will not require restoration to "show" quality. This program can be removed from your current projects list."

Then, on October 14, 1959, "Subject car consists of a fiberglass body with approximately thirty magnesium skin components." That would have been the mule body and all the spare body panels. Ironically, not a single piece of the extra body panels were ever used.

Also, "Confirming conversation with you on October 13, the above will be salvaged five days from this date unless instructions are received to the contrary." As far as we know, the mule body and the spare body parts were scrapped in mid-October of 1959.

In 1967 the Corvette SS was donated to the Indianapolis Motor Speedway Museum, where it is on display today.

World class driver Sterling Moss jumps from the mule to acclaim the handling of the mule to Duntov. At this point the mule has almost 2,000 test miles, and the handling and brakes have been fine tuned. Despite being overweight, it is turning laps around the 5.2 mile track that are four to six seconds off the track record. (Ludvigsen Library)

Friday, March 22, 1957, the Corvette SS arrives at last. It is taken to the technical inspection area for a safety check. Note the bubble top in place. Thanks to the technicians at styling, it has a show car finish. However, this perfect finish was one of the factors that delayed the car from being tested prior to the race.

The technical inspection of the SS continues. The bubble top has been removed, and the official in the background has the check list. Note the windshield wipers, which were added at the last minute to satisfy track regulations.

Next, it was photo session time. Here Duntov poses in the SS.

Then Corvette SS driver John Fitch, with the white-banded helmet, poses with the car.

Now recovered completely from his 1956 headache, Mauri Rose (right), Chevrolet engineer and three-time Indianapolis 500 winner, confers with Zora Arkus-Duntov (left) and John Fitch after a test drive in the SS.

Finally the Corvette SS gets on the track for a test run on late Friday afternoon. There are three problems: heat in the cockpit, a poorly placed inside rear view mirror, and locking brakes. In the evening the SS was modified, here we see insulation placed between the driver and the side exhaust, with a hole in the left door for air and a hose bringing in air from the right door. Also, a mirror has been added above the original mirror. (Bernard Cahier)

Last minute changes to the SS cockpit: air inlet hole in the driver's door, insulation around the driver's feet, an extra higher inside rear view mirror, and an air inlet hose which attaches to the passenger's door. (Bernard Cahier)

Saturday, March 23, 1957, Fitch is still having brake problems, but with no time left for more testing or adjustment, the SS was pushed to the starting line. Here John Fitch and Zora Arkus-Duntov discuss the situation, with the Corvette SR-2 and Corvette Number Three in the background. (Peter Brock)

Saturday, 9:59:55 a.m., the drivers are lined up in a Le Mans style start. Fitch is first in the Number One SS; next is O'Shea, the Number Two SR-2 driver; then Hawthorn, driving Corvette Number Three; and Thompson heading for Corvette Number Four.

Saturday, 10:00:01, they're off! First, a foot race to the cars, where the cars are started and driven away as quickly as possible.

Corvettes Number Three and Number Four are the first away from the start.

Shortly after the 10:00 a.m. start, the Corvette SS driven by John Fitch pulls out beside the 4.5 liter Number 19 Maserati driven by Jean Behra; teamed with Juan Fangio, Number 19 was the overall winner of the 1957 race. Number 20, a 3.0 liter Maserati, being driven by Sterling Moss, finished second overall. (Jerry McDermott)

11:30 a.m., the Corvette stalls out on the track. Fitch is able to coast into the pits. Valuable time is spent looking for a fuel injection problem, shown here. Finally, the coil is touched, and the car starts, Fitch is back in the race. (Bernard Cahier)

12:30 p.m., out on the track the coil fails again, but Fitch has a spare. After changing the coil, the SS runs strong and fast. Note the air pickup scoop on the passenger door.

1:05 p.m., Fitch pulls in for a routine driver change; Italian driver Piero Taruffi begins his turn at the wheel.

About 1:15 p.m., Taruffi only goes one lap, the suspension has failed, and the car is uncontrollable. The SS is considered unsafe and is withdrawn from the race. Duntov looks at the chassis in an attempt to locate the problem.

1:50 p.m., the SS sits in the pits after 23 laps, 119.6 miles, to finish 42nd in a field of 66 cars. It was officially withdrawn due to "overheating". A failed rear suspension bushing was found later. Note the air pickup "pod" on the driver's door. (Bernard Cahier)

After the Sebring race, the SS was being prepared for the next race — Le Mans — until the AMA racing ban halted the project. The SS was stripped of its name and prepared for various auto shows. Here it is shown at the Michigan State Fair, Aug. 30 thru Sept. 8, 1957. All racing modifications, including the driver's door air scoop and the extra inside rear view mirror were removed.

December 1958, Duntov prepares to drive the Corvette SS on the five mile circular track at the GM Proving Grounds in Arizona.

February 1959, the new Daytona Speedway is officially opened. One of the fastest first laps around the 2-1/2 mile track were turned by Duntov in the SS at 155 mph. (Daytona Racing Archives)

The Corvette SS was donated to the Indianapolis Speedway Museum in 1967. Here it poses beside the narrow brick strip at the Speedway's start/finish line. (Indianapolis Speedway Museum)

The Corvette SS was rebuilt in 1987 in preparation for the Historic Races at Laguna Seca in August. After a drive around the track, John Fitch sits in the driver's seat with Duntov as a passenger. Lou Cuttita, the engineer who restored the SS, is in the black cap, resting his arm on the SS, next to Duntov. In the background is Rich Mason, the current owner of the former Jerry Earl SR-2. (Noland Adams photo)

Fuel Injection

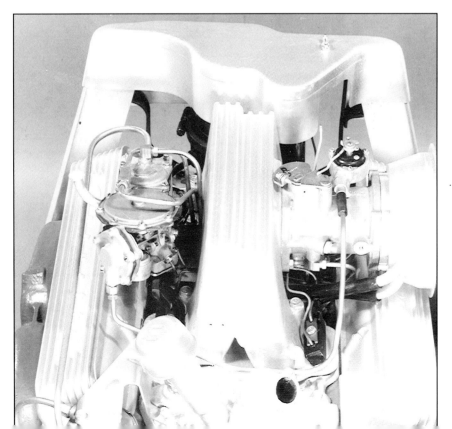

Chapter 6

Fuel Injection is a device used to introduce fuel into an engine. It's not a new system. Dr. Rudolph Diesel applied for a fuel injection patent in 1892. He was so successful that engines and the fuel they burn, bear his name, Diesel.

Most of us know that several GM automobile engines in the fifties and sixties had fuel injection. The standard method of combining fuel and air to be burned in an automobile engine was to use a carburetor. The use of fuel injection on an automobile engine met with mixed reaction, not everyone was pleased.

The General Motors fuel injection system's history can be traced back to the 1920s. Besides development for diesel engines, it was further refined for aircraft engines. By 1930 over 350 technical papers had been written about the use of fuel injection on aircraft engines.

In 1927 the Engineering Department of the Buick Motor Company was located in Flint, Michigan. The engine section hired a draftsman named John Dolza. Later, Dolza and his engineering team was responsible for many major developments. The two that deserve a close look are the casting method for the 1955 Chevrolet V-8 engine, and fuel injection for GM engines.

Zora Arkus-Duntov was a brilliant engineer who concentrated on the Corvette. John Dolza was also a brilliant engineer, but his career was spread throughout GM. In 1927, John Dolza was a young, well educated Italian engineer. He received a master's degree in both electrical and mechanical engineering from the Polytechnic Institute in Turin, Italy in 1926. While at the institute he designed the school's altitude laboratory.

This is a photo of GM Engineer John Dolza, the inventor of Rochester Fuel Injection, at age 42, taken in 1944.

In 1929 and 1930, Dolza worked under Buick's chief engineer, Dutch Bower. Dolza was instrumental in the design of the first Buick valve in head "straight eight" engine in 1931. Dutch Bower and John Dolza became close friends, and Dutch characterized Dolza as "the most brilliant engineer I've ever known or had the pleasure to work with."

In 1940 Mr. Dolza was transferred to GM's Allison Division in Indianapolis, where he was engaged as consulting engineer on special assignments. Major projects include developmental work on fuel injection and other automatic controls for aircraft engines, aircraft turbo automatic controls, pilot automatic controls, lubrication systems for high-altitude aircraft engines, and turbo-prop engines and their related controls.

In 1945 Mr. Dolza was transferred to the GM Engineering Staff in Michigan. By 1952, John Dolza had been promoted to the engineer in charge of the Power Development Section of the General Motors Engineering Staff. About this time, Mr. Dolza and his staff developed a thin wall, "green sand", engine block casting process. This engine design was a new, compact V-8 with 240 cubic inch displacement (cid).

This engine was available to all GM divisions, but it was untried. It was considered to be too small in displacement, the lightweight overhead valve train was unproven technology, and the thin wall iron block design required a breakthrough in casting technology for economical mass production.

At that time, early 1952, Ed Cole was the Works Manager for the Cadillac Division. Cole was promoted to Chief Engineer at Chevrolet Division. Cole soon learned that Dolza had perfected the thin wall, "green sand", casting process technology needed for the new small GM V-8 engine. Recognizing Chevrolet's need for more power, Cole resurrected the small GM V-8 for a second look.

Later, Cole got credit for recognizing a "jewel in the rough" in the undeveloped engine. Cole recruited Clayton Leech, a Pontiac engineer, to perfect the valve train. Thanks to Dolza's new casting process, the cid was increased from 204 to 265, producing 162 horsepower.

John Dolza had been observing fuel injection systems since his aircraft engine work in the early forties. He had traveled Europe extensively studying state of the art diesel fuel injection systems. Dolza also spent some time with Ferrari, who was already using fuel injection on its racing car engines.

In the late forties and early fifties, Cadillac was the division that led in technology, no doubt due to the presence of Ed Cole as Chief Engineer. Since Cadillac had an image of new technology, new engines, transmissions, suspensions, and other new systems, with few exceptions, were introduced on Cadillacs.

By 1953 two crude fuel injection systems for automobile engines had been developed within GM. Later in 1953 a separate department was formed to pursue research and development between the two different types of fuel injection systems. This was independent of Cole's development of the 265 cid V-8. At that time, the fuel injection system was probably intended for the Cadillac Division.

GM's system "A" was a continuous flow system, developed by John Dolza at the GM Tech Center, with joint efforts of GM's Corporate Engineering Staff and the AC Spark Plug Division.

GM's system "B" was a timed injection system, developed jointly by GM's Diesel Equipment Division and GM's Research and Development Staff. Both the "A" and "B" systems were in crude forms of development and testing in mid-1954. Each system had its own share of problems to be solved before they could even be considered for production.

Fuel injection was the buzzword around the Detroit automotive research and development community in 1954. The American Society of Mechanical Engineers (ASME) was only one of many technical societies to publish technical papers relative to fuel injection. Their paper titled, "Dynamics in the Inlet System of a Four Stroke Single Cylinder Engine," was published on Dec. 3, 1954.

After the successful introduction of the 265 cid V-8 engine at the start of 1955 production, Ed Cole turned to Dolza's "A" system fuel injection. Now Chevrolet was competing with Cadillac and Pontiac Divisions for the system.

In early 1955, Ed Cole assigned Zora Arkus-Duntov to investigate the application of the GM "A" system. After Cole left Cadillac in 1952, their aggressive pursuit of technology began to wane. By 1955, Cadillac was proceeding cautiously, and they backed away from the fuel injection leadership role in GM. This left Chevrolet to battle with Pontiac for the honor of GM's fuel injection debut.

Zora Arkus-Duntov immediately accelerated his research and development work on the GM "A" continuous flow fuel injection system. He tried it on a standard 1955 V-8 test engine with a four-barrel intake manifold. Unfortunately, the fuel injection showed no relative power advantage over the same engine with the stock Carter WCFB carburetor.

In mid-1955 Zora Arkus-Duntov got a 1955 Corvette for testing. It was fitted with a "prototype 1956 four-barrel Power Pack V-8." Next, the test Corvette was fitted with what was described as a "cobbled up 1955 four-barrel V-8 with fuel injection apparatus." Both Cole and Duntov were pleasantly surprised when the test track results showed an increase of almost 10 percent for the fuel injection system.

On Sept. 9, 1955, Zora Arkus-Duntov drove a 1956 Chevrolet prototype up Pike's Peak, setting a new record. This accomplishment put Chevrolet in a new light. The publicity was welcome, and new car sales were boosted by the new image.

Now, Mr. Duntov wanted to improve the image of the Corvette. He planned to prepare several Corvettes for racing and earn a much needed publicity boost. Mr. Duntov took his plan to Chevrolet's Chief Engineer Ed Cole for approval.

Cole considered the situation carefully. It was October of 1955, and engineering work on the 1956 Corvettes had been completed. Production of the 1957 Corvette was scheduled to begin in about 10 months. Ten months seemed like plenty of time, but Duntov was already working with Dolza to perfect the new fuel injection system.

Cole wanted the new fuel injection system to be de-bugged so it could be installed on new Chevrolets and Corvettes at the start of 1957 production. Duntov assured Cole that there was plenty of time to do both projects. Cole enthusiastically agreed to approve Duntov's plan.

In early November of 1955, a flow box operator named Lou Cuttita, was testing the new fuel injection units. He observed the major breakthrough that fuel injection needed. His persistent and exhaustive bench testing of fuel injection test mules revealed a new principle. "Dynamic supercharging, a ram air intake tube creates a pressure pulse at the intake valve."

Many different variations of fuel injection systems were tried before the system worked satisfactorily.

Even as late as December 1955, the two GM fuel injection systems were still being developed. Politics between divisions were intense, and a final decision over which system would be developed for production dragged on and on.

Harlow "Red" Curtice was President of General Motors. In his previous job, General Manager of the Buick Motor Division, he had approved construction of the 1938 Buick Y job, GM's first show car. As GM President, he had approved the 1953 Motorama Corvette in 1952. At the '53 Motorama, he announced that Corvette production would begin by the end of June, 1953. Red Curtice also approved the use of the Dolza inspired V-8 engine for the 1955 Chevrolets.

Harlow "Red" Curtice was going to make the decision on which fuel injection would be developed for GM production cars. Dolza's "A" system was adapted to several 1956 Chevrolets by Zora Arkus-Duntov. These test units had a large central plenum chamber (commonly called a "doghouse") with a separate air-meter, fuel-meter, and fuel-nozzle system.

The timed fuel injection unit, system "B", was experiencing timing problems. Plus, this more complex system was described as a "plumber's nightmare."

One of Red Curtice's considerations was adaptability. The "A" system, with its separate sub-assemblies, could be used and adapted to every engine configuration in GM's varied engine line-up. The "B" system, although backed by the mighty Detroit Diesel Division, posed more costly and difficult adaptations for some engines.

Another factor that cannot be overlooked goes back to the Buick Motor Division. Red Curtice was the General Manager, and one of the engineers he counted on was John Dolza. At that time, Ed Cole was Chief Engineer at Cadillac. Meanwhile, Zora Arkus-Duntov was improving the Corvette.

Now it was late January of 1956, and time for Red Curtice to approve one of the fuel injection systems. Curtice had a personal association and confidence in Ed Cole and John Dolza. Having Zora Arkus-Duntov also seemed like the perfect team; so Curtice selected the continuous flow "A" system.

Rochester Products Division began the GM Fuel Injection "CRASH" Program in February, 1956. The first batch of Prototype fuel injection units were available in June, 1956, and were mated with prototype Chevrolet 283 V-8's for test vehicles at the GM Technical Center in Warren, Michigan.

Chevrolet began preparing the St. Louis assembly plant for the 1957 Corvette pilot car run during the production of 1956 Corvettes. Pilot cars are advance models of next year's production. These pilot runs are supposed to determine if the parts supplied for the next model year will fit as a unit, and the assembly line will be checked to see if any modifications are needed to the existing equipment.

The 1957 Chevrolet 283 cid V-8 engine pilot run began on August 6, 1956 at the Flint Motor assembly line. The Rochester fuel injection units were not ready, so no fuel injected engines were included in the pilot engine build for shipment to St. Louis.

The actual build date of the 1957 pilot line at the St. Louis Corvette assembly plant began on Monday, August 13, 1956. Because the 1956 and 1957 bodies were identical at this point, no special equipment was needed for vehicle assembly. However a few test stands were converted to be able to test the fuel injected engine and transmission assemblies. The total quantity of pilot line Corvettes was probably between four and 10 units, with various carburetor systems and transmissions.

Some fuel injection units being test run on engines.

The Flint Motor assembly line was directed to assemble its first group of 1957 fuel injection engines. The date was August 29, 1956, with the engine assembly code being F829X. This code means, F = Flint Motor; 829 = August 29; X = Experimental. This is because the fuel injected engines had not been assigned an engine suffix code yet. This was a small batch of engines, probably less than ten in all. They were shipped to the Chevrolet Engineering Center at the GM Technical Center in Warren, Michigan for intense and torturous, but successful, dyno testing.

The Flint Motor assembly line was approved to begin the Corvette fuel injection pilot engines on Monday, September 24, 1956. The St. Louis Corvette assembly plant began building 1957 Corvettes on Monday, October 1, 1956 and began including the pilot fuel injected Corvette engines almost immediately. The early serial numbers on some fuel injected 1957 Corvettes suggest that perhaps as many as 12 early fuel injection units may have been shipped to the St. Louis assembly line for installation in a production Corvette.

By mid-October the Rochester Products Plant was ready to start building fuel injection units for Chevrolet. The first "Production" unit was ready on October 22, 1956. It was carefully placed in its shipping box by a gathering of Rochester officials. The fuel injection units were to be installed on the engines at the Flint Motor assembly plant, so they were shipped from Rochester, New York to Flint, Michigan.

At first, the fuel injected engines for production cars were installed only on Corvettes. As the supply became a bit more plentiful, they were installed on Chevrolet passenger cars as well. The fuel meter had provisions for mounting the high pressure fuel pump on the side, as required for Pontiac engines. So there were a few fuel injected Pontiacs, too.

The Rochester Ram-Jet fuel injection went on to further development. It was soon dropped by Pontiac, and Chevrolet passenger cars dropped it early in the 1959 model year. Other automobile manufacturers turned to supercharging or multiple carburetion, while Corvette stayed with the constant flow fuel injection until 1965. By then carburetors had improved, and they replaced the more complex fuel injection system.

It was many years later when electronic fuel injection was developed. Ironically, most gasoline burning internal combustion engines use fuel injection now.

Actually, we could have gone further with the Rochester fuel injection development. But we achieved our goal, to present the history of the development of the system.

One experimental fuel injection system was adapted to a two barrel carburetor intake manifold. The engine ran, but there was no increase in power over carbureted engines.

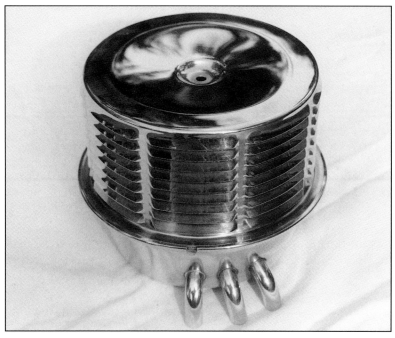

This air cleaner was developed for the Rochester fuel injection system. The general feeling is that this unit restricted air flow. One must examine the air cleaners used with two four-barrel carburetors, each has four rows of louvers. This air cleaner has the eight rows of louvers, plus an extra row, nine rows of louvers in all. Unless the element was full of foreign material, the air cleaner probably was not the problem. However, it is a sealed unit, the element cannot be changed.

A breakthrough for the Dolza designed fuel injection system occurred when flow box operator Lou Cuttita correctly identified air flow patterns.

An early fuel injection unit on a 1956 Chevrolet passenger car at Rochester Products. Note the louvered Corvette air cleaner. (Elmer F. "Pete" Detiere)

October 22, 1956, Rochester Products, Rochester, N.Y., "Job One," the first 7014360 Rochester Fuel Injection unit, is ready to ship. Left to right, E.A. Kehoe, Chief Engineer; E.F. Detiere, Supervisor of Engineering Testing; J. Mathews, Production Manager; H. Sutton, Foreman, Fuel Injection, (placing the unit in the box); H. Brandt, General Manager, Rochester Division; H. Grassmick, Superintendent, Fuel Injection; C. Brandon, Director of Production Control. (Elmer F. "Pete" Detiere)

One of the first Corvettes to get an experimental fuel injection unit was this 1956 Corvette, shown at the Tech Center in Warren, Mich.

This is probably the first 1957 Corvette to get an experimental fuel injected engine. Note the flag emblem in the cove. The fuel injection emblem was not yet available.

A production fuel injected engine on a display stand.

The second type of fuel injection air cleaner with a replaceable element.

"It's here! Chevrolet's Ramjet Fuel Injection!" Images of a cut-away fuel injected engine are sent to TV viewers via CBS during the 1957 Chicago Auto Show. (AMA)

A fuel injected 1957 Pontiac at a 1957 Auto Show. (AMA)

This common looking box was discovered at a swap meet. Inside? A brand new Rochester fuel injection unit, keep your eyes open!

Auto shows

Chapter 7

The 1957 Corvette Super Sport in a GM publicity photo.

If this were 1956, this chapter would be entitled "1956 General Motors Motorama." However, there was no Motorama in 1957. Overall sales of passenger cars in 1956 were down, causing GM upper management to reconsider all major expenditures. As a result, the 1957 GM Motorama was never funded or developed.

The GM Motorama, a touring group consisting of one-of-a-kind dream cars, current GM production cars, and show biz, certainly drew large crowds. But it was quite costly. Designing and building special show cars, designing and building the platforms and displays, renting large buildings in major cities, and moving all the cars and equipment from city to city was expensive.

The 1953 to 1956 yearly Motoramas were justified because of their impact on new car sales. The Motoramas clearly influenced purchasers, who bought the GM car of their choice. But 1956 was different. In spite of the Motorama, the sales of GM passenger cars were down. There seemed to be no choice. There would be no GM Motorama for 1957.

However, GM's new 1957 models could still be seen, but on a different scale. All major cities had an auto show where new car models were previewed. The problem facing GM was, now there would be all makes and models present. Before, the Motorama was a showcase for just GM cars. Now there was competition from other car manufacturers.

But GM had no dazzling special dream cars like those that had been built for the Motoramas. That budget had been axed, but GM still needed to present its cars in a favorable light, within a tight budget.

There were probably other GM cars specially prepared for the 1957 auto shows. The car we are concerned with started out as a stock 1956 Corvette. On the surface, the show car appears to be a mildly modified fuel injected 1957 Corvette. The obvious differences are the small windshields, racing stripes down the middle, and a couple of shiny side panels with small scoops. But it was much more than that.

First, this 1957 show car was named the Corvette Super Sport. The problem was that Corvette had two other Corvette Super Sports that got a lot of attention from the automotive press. The first SS was the mule, based on a race car chassis with a fiberglass body, it was a crude test car. That car had been called the "white mule," "white SS," and "Corvette Mule."

The second SS was the actual Corvette SS (Super Sport) racecar. This car was developed using experience gained from the SS mule. The racing SS had a special fuel injected engine with a magnesium body, and it did race at the 1957 Sebring race.

(Right) The main floor of a 1957 auto show. The Corvette Super Sport is in the center.

The Corvette Super Sport amid admirers on the main floor of the auto show; we were told this was in Chicago.

THE CORVETTE SS SHOW CAR

The third Corvette SS was the stock Corvette modified for the 1957 show circuit. It was modified in December of 1956 and shown at the Waldorf-Astoria Hotel in New York City in January of 1957.

Although we've called it the third Corvette Super Sport, it may have been the first. The Corvette SS racing car and its mule test car were planned as far back as August of 1956, well before the Super Sport show car. However, it was referred to within GM as the XP-64, and it was finished in march, well after the show car had been on the show circuit for a couple of months.

We expect this was the order of events: 1) August 1956, XP-64 project is approved; 2) Early December, XP-64 mule is completed; 3) December, 1956, Super sport is built as a show car; 4) January 1957, Super Sport begins the show circuit; 5) February, 1957, the XP-64 is named the Corvette SS, with the understanding that it means Super Sport; 6) March, 1957, the SS is raced at Sebring. If true, why were two different Corvettes named SS or Super Sport? One must recall that the XP-64 racing car was a secret project, and the managers of one project may not have been aware of the other.

Besides being shown at the New York Auto Show, it made a regular circuit of many of the 1957 auto shows. Its full schedule is not clear, but it was also shown in Chicago in February of 1957.

The Super sport has several obvious external features. There are three stripes the length of the car, with the widest being in the center. The large stock windshield is replaced by two very attractive smaller bubble windscreens. The tires have narrow whitewalls. The side coves are covered with sheet metal, with a reverse scoop in the rear of the cove.

Most of the work was done to the interior. The door panels, seats, carpets, pedals, steering wheel, and dash are all special for this car. In addition, there is special lighting, the courtesy lights are mounted in the door and shine on the ground, not in the car.

The car was wrecked in the left front many years ago. Fortunately, the car still has all of its unique parts and features. It is in private hands and is being repaired and restored to original show car condition.

Interior of the Super Sport shows its unique features. (Jerry McDermott)

CORVETTE: AMERICAN LEGEND

Daytona

Chapter 8

Shortly after the turn of the century, kids of all ages with high-powered cars were driving them at high speeds on the sands at Daytona Beach, Florida. By February 1957 the event had become one of national importance. The official name was the NASCAR Speed Weeks Performance Trials. There were one-mile standing start acceleration tests, and one-mile top speed tests that averaged the speed for two runs from opposite directions.

Circular track racing was held on a 4.1 mile course that was built shortly after World War II. One stretch was on the beach, the other long straightaway was on the paved highway. New residences and motels were being built along the highway, and enough space for the races was getting hard to find. An all-new track was planned for 1959.

Several Chevrolets participated in the runs on the beach. Super mechanic, Smokey Yunick, taped off the openings on a stock Cameo pickup and turned 132.353 mph.

Bill Mitchell's SR-2 was there, complete with a red and white striped rudder, like an old fighter plane. *Corvette, America's Star Spangled Sports Car*, described the SR-2, "its extended nose and extra-light nacelles under the headlights were like those of the first SR-2. A large finned headrest was added, striped in red and white like the rest of the car. For Daytona it was fitted with a Plexiglas enclosed canopy, skirts for the rear wheels, and Moon discs for the front ones, fairing cones for the front lamps and Plexiglas covers for the sides of the grille."

Buck Baker drove the SR-2 to a standing mile average speed of 93.047, which won its modified class. Baker turned the flying mile with a two way average of 152.866 miles per hour, second best to a D-type Jaguar.

A team of modified Ford Thunderbirds were factory sponsored. The bodies were lightened and streamlined, including headrests. The engines were fuel injected and supercharged and coupled to a four-speed transmission, and the engines were placed several inches back in the chassis. Danny Eames won the experimental class with a standing mile run of 98.065. He was reported to have been near 160 mph at the end of the measured mile. The experimental Thunderbird posted a two way flying mile average of 169.811.

Other Corvettes dominated the production sports car class with three to five liter engines. The 283 cid engine is 4.6 liters. In the standing mile acceleration runs, Paul Goldsmith won the class with an average of 91.301 mph. The acceleration runs are averaged over the length of the course. In order to produce an average of 80 mph, the car must turn a bit over 100 at the end of the measured mile. Johnny Beauchamp was second with an average of 89.798, and Betty Skelton was third at 87.400 mph.

In the production sports car class flying mile, Corvettes dominated again. Paul Goldsmith won with the two way average of 131.941. Betty Skelton was second, and Johnny Beauchamp was third.

From the 1958 issue of *Motorspeed*: "Emphasis on sports cars was more extensive at the 1957 speed weeks, with a full scale racing program scheduled for the second Sunday at the nearby New Smyrna Beach Airport." The plan was to hold a pro-am race, with pros and amateurs competing. "Sports Car Club of America (SCCA) ruled its members could compete, although the events were non-sanctioned.

Bill Mitchell poses with his SR-2 in the Michigan snow just before it was shipped to Florida for the races at Daytona Beach.

"SCCA's champion Carroll Shelby, in a Ferrari, was the feature race winner, with Marvin Pance, a NASCAR stock car professional, second in a Ford Thunderbird.

From *Corvette, America's Star Spangled Sports Car*: "Potent factory-backed Ford Thunderbirds were threatening strong opposition, so the Corvette entries were strengthened by bringing over one of the cars that was being tested at Sebring for the twelve-hour race. Although beaten by O'Shea's 300SL in the production race, Paul Goldsmith's Corvette, qualified third fastest behind two Ferraris for the main event and had a magnificent battle for second with Marvin Panch's T-Bird, in the race until Paul had to pit for a fresh tire. The former motorcycle racer placed fourth."

Mitchell's SR-2 was the pace car for the race. There were several minor accidents and one serious injury. Mike Marshall, a Miami sports car dealer, was critically hurt when his Porsche Carrera flipped in practice. He was hospitalized and under treatment for many months.

Mitchell's SR-2 arrives from Michigan on an open trailer. Or, did it just come from Smokey Yunick's garage across town?

The Corvettes get a little tuning attention. Note the air cleaner on the fuel injected Corvette in the foreground. The dark Corvette directly behind is carbureted, as its two air cleaners are visible.

As the cars are lined up for the public to view, the SR-2 gets its share of attention.

CORVETTE: AMERICAN LEGEND

A team of streamlined Thunderbirds, factory supported, were entered in the 1957 Daytona Beach races. The fuel injected and supercharged cars had their engines set back several inches, they were lightened, and had four-speed transmissions.

The cars line up for the race; besides Corvettes, we see a Thunderbird, Jaguar, Mercedes, Porsche, and others.

Betty Skelton makes a low speed publicity run in the SR-2.

Number Nine Corvette takes the flag at the end of a high speed run.

Betty Skelton, driving a special Corvette, script in the side cove says "Corvette," at the airport entrance. (Daytona Racing Archives)

The SR-2 was the pace car at the New Smyrna Beach races. (Bill Tower)

Chapter 9

Sebring

The week of March 18, 1957 was a week of shakedown and test runs for the sport cars entered in Saturday's Sebring race. The Corvette SS will not arrive until late Thursday night, so the test SS, called the mule, is shown here as a practice car. (Bernard Cahier)

The Sebring Airport was first recommended for use as a racing circuit by Eric Ulmann. He helped stage the Watkins Glen Road Race in 1948 and 1949, and the Palm Shores Road Race in January 1950. Ulmann longed for an American endurance road race of major importance like the Le Mans race in France.

Sebring had a generally favorable climate, but it was in a remote location. It had been used as a B-17 training base during World War II, but it was seldom used in 1950. The first race at the Sebring airport was held December 31, 1950 on the taxiways, side streets, and runways near abandoned military barracks, aging aircraft, and dilapidated hangers and warehouses.

By 1957 the Sebring race had expanded to a 12-hour event of major importance. Its official title was the Florida International 12 hour Grand Prix for the Amoco Trophy. Juan Manual Fangio and Jean Behra co-drove the new 4.5-liter Maserati to win the 1957 race. Other important entries came from Jaguar, Ferrari, Alfa Romeo, Renault, Lotus, Cooper, Porsche, OSCA, A.C., Austin-Healey, M.G., Arnolt-Bristol, Morgan, Triumph, and — Chevrolet Corvettes. 66 entries in total.

Chevrolet backed the Corvette SS, although it was technically entered by Lindsay Hopkins of Miami, Florida. The Corvette SS, wearing Number One, was a special racing car, which was barely completed before this race. The cars with the largest displacement engines started in front. The SS was driven by John Fitch and Italian Piero Taruffi, but it had mechanical problems and did not finish the race.

Several other Corvettes were entered, with two being factory sponsored. Number Four was entered by John Fitch, and was driven by Dick Thompson and Gaston Andrey to a 12th place finish. Number Four completed 173 laps, winning the GT class. It was fuel injected, like all the Corvettes entered.

The other stock appearing Corvette, Number Three, was driven by Dale Duncan, John Kilborn, and Jim Jeffords. It completed 168 laps to place 15th, the winning Maserati completed 197 laps.

While the Number Three and Number Four Corvettes looked stock, they were not. Both were fuel injected, with cast iron four-speed T-10 transmissions, experimental at this point. Both cars were 1957 models built in November of 1956. Many new performance items were built into the cars. For a shakedown, they were taken to the Bahamas for the Nassau Trophy Races starting on December 7. While their performances were not considered a great success, a number of small problems were uncovered, like fuel injection surging, overheating, handling, and braking. Overheating was traced to bad head gaskets, which was soon cured. The fuel injection mechanical parts were redesigned.

Both stock bodied cars arrived at the Sebring track early, along with an extra practice car. *Corvette, America's Star Spangled Sports Car*, quoted Dick Thompson, "We experimented with them at Sebring and for the first time since I started racing in 1953, I had my fill of driving. I would take one Corvette out and pound it until something gave way. Then I would trade it immediately for the other one. It was fatiguing but fascinating." It is to the credit of all that both Corvettes survived and finished the race.

Repairs were started on the four factory-backed Corvettes immediately after the Sebring race. This included the Corvette SS, the two fuel injected cars Number Three and Number Four, and the SR-2. Although the SR-2 was owned by Bill Mitchell and entered in the Sebring race by Lindsay Hopkins of Miami, it was considered to be a factory-backed car. When the AMA ban began to take effect in late April, all GM divisions ended their involvement in automobile racing.

The two stock Corvettes and the SR-2 were eventually sold to private owners in the summer or fall of 1957. Private owners continued to race their Corvettes at races like the Milwaukee SCCA in April of 1957 and Road America in June. The factory backed Corvettes were withdrawn completely after the Sebring race.

World-class driver Sterling Moss jumps from the mule SS to congratulate Duntov on the car's performance. (Ludvigsen Library)

Friday, March 22, 1957: The Corvette SS arrives, and is taken to the tech inspection area with the bubble top in place. It has a show car finish, one of the factors that delayed its arrival in Florida.

Lined up for tech inspection, we see the two factory backed Corvettes with rows of spectators looking on.

CORVETTE: AMERICAN LEGEND

Another view of the tech inspection area, with the SR-2 Corvette, Number Two, in the background.

The stock bodied practice Corvette being tested on the track before the race.

Duntov poses in the freshly painted and detailed Corvette SS.

Corvette Number Three and the practice car, numbered P3, sit in the pits during practice.

The SR-2 poses with its mechanics and drivers during a practice session.

Now recovered completely from his 1956 headache, Mauri Rose (right), Chevrolet engineer and three-time Indianapolis 500 winner, confers with Zora Arkus-Duntov (left) and John Fitch after a test drive in the SS.

10:00:01 a.m., they're off! First, a foot race to the cars, where the driver jumps in, fastens the seat belt, starts the car and drives away as quickly as possible.

Corvettes Number Three and Number Four are away first.

The Corvette SS driven by John Fitch pulls out beside the 4.5 liter Number 19 Maserati driven by Jean Behra, the overall winner; Sterling Moss is driving Maserati Number 20, which finished 2nd overall. (Jerry McDermott)

The SR-2, getting attention in the background, suffered a number of mechanical problems, while the Number Four Corvette gets its gas tank filled.

The Number Three Corvette sits in the pits ahead of the SS, which was plagued by a number of problems. Tires, overheating, fuel injection, and ignition were some of the problems.

The Number Three Corvette beside a row of old warehouses.

CORVETTE: AMERICAN LEGEND

After solving all the previous problems, the SS is now turning fast laps. But a suspension part failed shortly after this photo, and it was out on lap 23. Here the SS driven by Fitch chases the D-type Jaguar of Ensley/O'Conner, who in turn chases the Number Four Corvette of Audrey/Thompson.

About noon, the Number Four Corvette passes the start/finish line.

CORVETTE: AMERICAN LEGEND

The SR-2 stops for Amoco fuel, the filler neck is in the headrest, and an underhood check.

A tire wear and temperature check.

And it's away again. Those are track officials in the funny hats.

Number Four comes in for left side tires and fuel.

And rushes back on the track.

The SR-2 is still running strong.

The Number Four Corvette has an overheating problem in the early afternoon, helped by a quick shower. Note the right side fuel filler cap, and the colored marker lights on the hardtop.

Number Three is still running strong.

The SR-2 is still in the race, too.

The pit crew sends a note to Thompson.

Flash cameras are not allowed in the pits after dark, so our next shot is the SR-2 taking the checkered flag!

Ed Cole congratulates the driver, Lovely or O'Shea.

The AMA ban stopped factory support, but here a couple of Corvettes battle it out with a herd of Jaguars at the Milwaukee SCCA races in April of 1957. (Jerry McDermott)

At Road America in June, 1957, entries included a stock bodied Corvette and the SR-2 belonging to Jerry Earl, now with a high fin. (Jerry McDermott)

4-Speed Transmission

Chapter 10

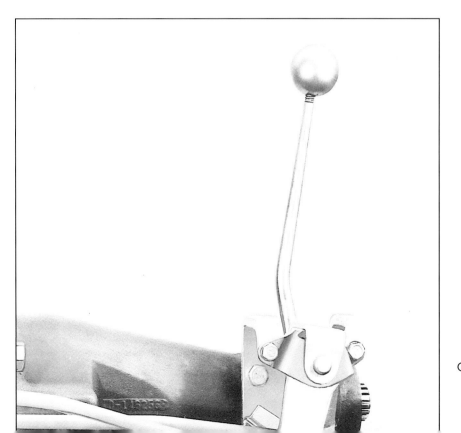

From the beginning of Corvette production in 1953, the two-speed automatic Powerglide transmission was used in the Corvette. The reason was simple, a manual transmission that could handle the extra torque of the Corvette engine was not available.

This lack of a strong manual transmission was emphasized when the 195 hp V-8 was available for the 1955 Corvette. A three-speed manual had been developed in 1954 by Saginaw gear for use in 1955 Chevrolet passenger cars. But it was mid-1955 production before the three-speed was modified and installed in Corvettes on an experimental basis. By the end of 1955 Corvette production, as many as 40 Corvettes, had three speeds installed by the factory.

In 1956 the three-speed transmission became a standard item, with Powerglide an option. These three speeds proved their worth both on the street and the track.

By 1957 the engine size was increased from 265 to 283 cid, with up to 270 horsepower. But a four-speed transmission was needed to keep the engine in a good torque range regardless of vehicle speed. Borg-Warner adapted a truck three-speed transmission by relocating the reverse gear into the tailshaft section, and replacing it in the main case with a fourth gear. This transmission was designated the T-10.

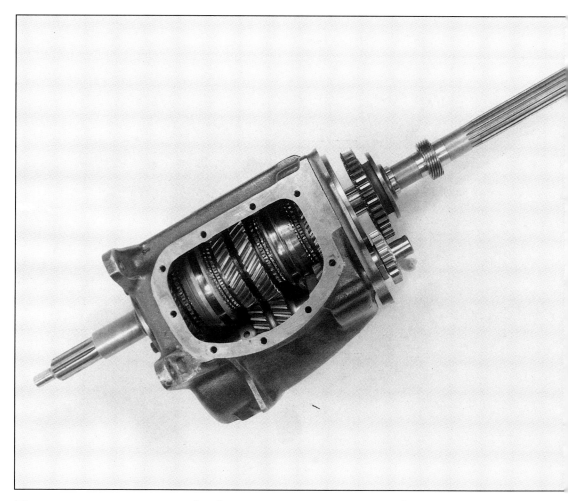

The gear arrangement is the Borg-Warner T-10, showing how the reverse gear was relocated to the tailshaft section.

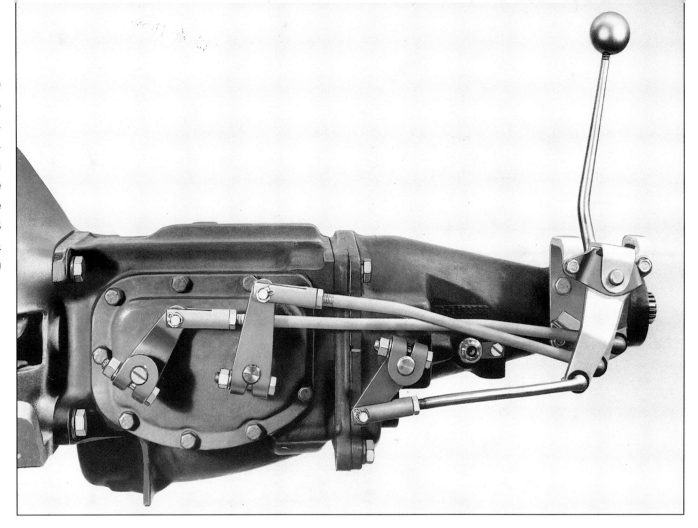

A 1957 T-10 with the side covers, shifting rods, and shifter in place. Note how the shifter rods cross. (Road & Track)

Fuel injected engines producing 283 hp were introduced on Corvettes shortly after the start of 1957 production. The T-10 transmission was able to handle the extra power and torque, so its availability as an option on corvettes was a prime goal.

The lack of parts kept the T-10 transmission from becoming an option on early 1957 Corvettes. The transmission was first available in the Assembly Instruction Manual (AIM) on April 9, 1957. The earliest known 1957 Corvette with a factory installed T-10 is serial number E57S103567 (number 3,567).

The main case was cast iron, with a cast aluminum tailshaft housing. An all aluminum transmission would have been much lighter, but the cast iron main case was cheaper. It was many more years before T-10s used an aluminum main case.

The installation of T-10 transmissions in early 1957 Corvettes is also possible. Early experimental four speeds may have been installed by the St. Louis assembly line. This early Corvette with a T-10 would have been shipped to the GM Tech Center or the GM Proving Grounds for evaluation. If it passed the engineer's tests, it would be sold as a used vehicle. Sometimes it went to a nearby Chevrolet dealer, or it might have been purchased by a GM employee.

There is another reason for an early 1957 Corvette to have a T-10 transmission. After the demand at the assembly plant was satisfied, T-10s could be purchased over the parts counter at Chevrolet dealers. They came as a complete changeover kit. So complete, in fact, that it is impossible to determine if the T-10 is original to the car, or is a retrofit kit. Many original castings were not dated, but if they are, one can determine when the transmission main case or tailshaft housing was cast. If they were cast after the car was built, it was not original to that car.

T-10s in 1957 Corvettes became hot items, because literally everyone wanted one. T-10s were used in Corvettes from mid-1957 through mid-1963.

The right side of a 1957 T-10 transmission shows the location of the casting numbers and dates. The main case is cast iron, with an aluminum tailshaft housing.

Body Modifications

Chapter 11

CORVETTE: AMERICAN LEGEND 251

There was only one major modification to the 1957 Corvette body, and it was added during the production year. It was discovered that the cowl area and the doors of older Corvettes were sagging after a few years.

In the cowl area, distortion of the body panels was caused by the weight of the rather heavy windshield frame. Some of the 1953 to 1956 Corvettes have such a problem in this area that extensive body repairs are required. Without a proper position for the windshield frame, the folding top cannot line up properly. On 1956 and early 1957 Corvettes, the front edge of the window door post may not seal against the windshield frame.

The problem with the doors is alignment of the latches. This can get so bad that the doors must be lifted to line up the latches just to close the door. If the hinge pins are also worn, closing the door can be a major chore even for a weightlifter.

The cowl braces may be seen on this upside-down front body section.

The special door bracing is shown here.

Cowl and door bracing in the form of specially formed aluminum channels was added in mid-1957 production. The date shown on the Assembly Instruction Manual (AIM) is July 11, 1957. The earliest Corvette reporting cowl and door braces was serial number E57S104296.

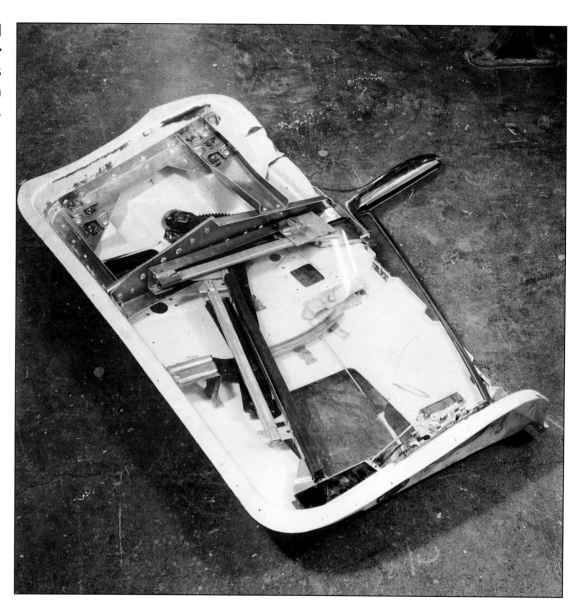

Mid-Year Performance Changes

Chapter 12

Chevrolet's involvement in racing can be traced back to its beginning in 1955. The credit for Chevrolet's successes in racing must go to racecar driver and engineer, Zora Arkus-Duntov.

While the 1956 Corvette's handling was adequate for everyday use on the street, several shortcomings kept it from performing well on a sports car track. The biggest problem was brakes, which were simply not up to the job.

The brakes in 1956 were the standard internal expanding drum type. This means shoes with special linings are forced against the rotating outer drums. The force required to slow the vehicle drum, being cast iron, expands away from the shoes, losing contact between the shoes and the drum, an effect called brake fade. After the brakes cool and the drum returns to its normal diameter, the brakes work again.

Brake fade can be reduced by several methods. Fins may be cast on the outer edges of the drum, which transfer heat to the air passing by the wheels and brake drums. This keeps the brakes cooler, delaying brake fade, and helps the brakes to cool faster.

Along with finned drums, cooling air may be ducted form the front of the car towards the brakes. In the front this is a relatively short path and it is the front brakes that do most of the work in stopping the vehicle. However, the rear brakes must be cooled, too. This is accomplished by ducting from the radiator area through the engine compartment, down to the rocker panels, exiting the rocker panels in an "S" shaped duct that directs the air towards the rear wheel.

The suspension was stiffened up and handling improved with new and bigger shocks front and rear, heavier springs front and rear, a larger front stabilizer bar, and a quick steering adapter. All of the above heavy-duty items were developed for the 1956 race at Sebring, and refined for the 1957 Sebring race. Shortly after, they were released as mid-year production options for the 1957 Corvette.

Other additional items were released for the 1957 Corvette in mid-production. One was wide wheels, permitting a tire with a larger "footprint." Production tires were skinny by today's standards: the wheel was 15" in diameter, but only 5" wide. This was one of the basic industry-wide tire sizes at the time. The optional Corvette wheel was 5-1/2 inches wide, permitting a special wider tire. However, this wheel did not have the wheelcover mounting "nubs" around the edge of the wheel. That meant the smaller, standard passenger car hubcap was used with the wider wheels.

Another 1957 option was the cold air box for fuel injected Corvettes. This fiberglass "box" was mounted on the left inner fender panel. It directed fresh air from beside the radiator back to the air intake of the fuel injection unit. Instead of consuming air that had been heated by its passage through the radiator, it was now burning fresh, "cold" air. The result was a denser charge, which expanded more in the engine. In some cases, the gain by the use of the air box was reported to be as much as 20 hp.

Yet another mid-year addition was the limited slip rear differential; GM's trade name was "Positraction." Using a series of internal clutches, a slipping wheel would transfer the power to the other side. Traction was improved because uncontrollable wheelspin was eliminated.

Combine all these goodies with the new four-speed transmission and fuel injection, and late production 1957 Corvettes had race-winning potential. However, due to the AMA racing ban, factory sponsored racing was dropped. Even Duntov's hands were tied. The heavy-duty items continued to be supplied to the private owners who chose to race their cars. But without further development and little support, heavy-duty options faded away rather than being improved upon.

Air was collected for front brake cooling by rubberized canvas scoops attached to the front backing plates. The scoop had to be flexible, to avoid interference with the front suspension during turns. The nickname for this large flexible scoop is "elephant ears."

Here is the underhood appearance of a 1957 fuel injected Corvette with ducts carrying cool air to the rear brakes. Note that the air cleaner element is visible, this is not an air box-car.

The right rear brake of a 1957 Corvette with ducting. Note the tube at the right carrying air from the rocker panel, the scoop on the backing plate, and the segmented brake shoes. With the drum and tire installed and the weight of the car on the tire, the hose and the scoop line up.

Looking from the rear of a chassis, with the body removed, we can see the edges of the left rear finned brake drum and the metal scoop that bolts to the backing plate.

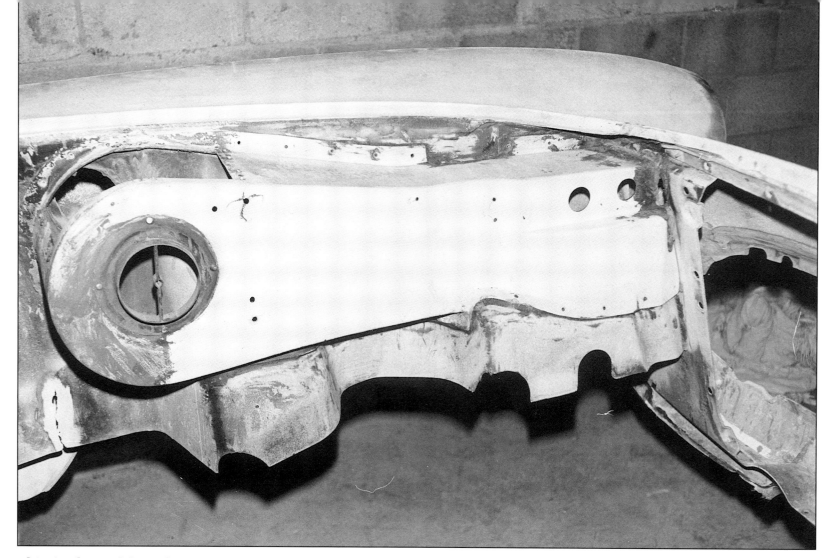

This is the cold air box, which supplied fresh air to the fuel-injected engine shown on a '57 Corvette under restoration. Air enters from the right, and the air filter is mounted inside the round hole on the left. The rectangular opening has a cover, screwed in place. The air box and the cover are fiberglass.

1957 Corvettes with wide wheels used this small passenger car hub cap.

Chapter 13
Corvette media origins

JUNE/JULY 1979

Corvette News was a quarterly publication for serious Corvette enthusiasts. Free issues were mailed to Corvette owners and others who filled out a card at the local Chevrolet dealership.

Joe Pike was an employee of the Chevrolet Motor Division. He met Ed Cole at the Pike's Peak races, and Cole brought Pike to Detroit. Pike was a serious Corvette enthusiast with the title, "Assistant Merchandising Manager." His real title was, "Head of Corvette Publicity."

Pike saw the need to publicize new Corvettes. At his urging, Chevrolet provided a publication titled *Corvette News* free to all Corvette enthusiasts. Postcards were sent to all Chevrolet dealers offering a Corvette owner's kit. The 1957 Corvette owner's manual also included a mail-in card with the same offer. The kit contained a patch, a lapel pin, an ownership card, and a subscription to *Corvette News*.

The first issue, Volume 1, Number 1, was published in the summer of 1957. The index page gave this information:

"THE *CORVETTE NEWS* BRINGS THE SPORTS CAR WORLD TO YOUR DOOR."

"Here is the first issue of Corvette News *- a magazine for owners of Chevrolet Corvettes. The* Corvette News *is designed to give you sports car competition results, articles about Corvette owners, technical information about the Corvette and other feature stories from the exciting world of sports cars."*

The cover of Vol. 1, No. 1 featured a shot of Corvettes racing at Road America. Subjects inside were: "Hot Corvettes Win 1-2-3 at Daytona Beach; Corvette Blazes to Victory at New Smyrna; Corvette Introduces Ramjet Fuel Injection at Nassau; Corvette and Owner Dick Thompson Win '56 SCCA Championship; New Corvette performance Features."

Corvette News, full of photos and technical information, was the boost Corvette sales needed. *Corvette News* was timely and well received, and it continued for many years. Once free, those old copies can bring Big Bucks.

Joe Pike was named "Head of Corvette Publicity" by Chevrolet General Manager, Ed Cole.